Freeway

Englisch für berufliche Schulen

Allgemeine Ausgabe

von
Rosemary King
Wolfgang Rosenkranz
Graham Tucker
und Oliver Frank

Ernst Klett Schulbuchverlage
Stuttgart • Leipzig

Freeway – Allgemeine Ausgabe

Englisch für berufliche Schulen

von
Rosemary King
Wolfgang Rosenkranz
Graham Tucker
und Oliver Frank

Begleitmaterial:
Lehrerhandbuch mit Audio-CDs (2) und Service-CD-ROM, 3-12-809941-3
Workbook 11/12, 3-12-809944-8

1. Auflage 1 7 6 5 4 3 | 2011 11 10 09 08

Alle Drucke dieser Auflage können im Unterricht nebeneinander benutzt werden, sie sind untereinander unverändert. Die letzte Zahl bezeichnet das Jahr dieses Druckes.

Internetadresse: http://www.klett.de

Redaktion: Volker Wendland, Astrid Keller (Assistenz)
Gestaltung: types GmbH, Stuttgart;
Thomas Gremmelspacher, Ulrike Wollenberg
Umschlaggestaltung: Nikolaus Keller
Grafik: Christian Dekelver, Weinstadt
Reproduktion: Meyle + Müller Medien-Management, Pforzheim
Druck: Himmer AG, Augsburg
Printed in Germany.
ISBN-13 978-3-12-809947-7

9 783128 099477

Freeway

Freeway – Allgemeine Ausgabe – für den Englischunterricht an berufsbildenden Schulen führt aufbauend auf der Fachoberschulreife zur Fachhochschulreife. Es entspricht den neuesten Richtlinien und Lehrplänen für berufliche Schulen der verschiedenen Bundesländer und orientiert sich zudem am Gemeinsamen Europäischen Referenzrahmen.

Freeway – Allgemeine Ausgabe – ist insbesondere an Fachoberschulen bzw. Berufsoberschulen und Höheren Berufsfachschulen sowie in entsprechenden Kursen an Fachschulen und Berufsschulen einsetzbar.

Freeway – Allgemeine Ausgabe – ist wie folgt aufgebaut:

Units 1 - 5	**Basic structures**
Units 6 - 10	**Advanced structures**
Units 11 - 12	**Special topics: Intercultural studies**
Units 13 - 14	**Special topics: World of business**
Units 15 - 24	**Exam preparation**

In den Units 1 – 5 wird dem unterschiedlichen Kenntnisstand und den verschiedenen Voraussetzungen bei Schülerinnen und Schülern der angesprochenen Schultypen durch eine flache Progression Rechnung getragen. Damit wird auch schwächeren Lernenden der Zugang zu den Themen erleichtert.

In den Units 6 – 10 werden Aufbaustrukturen vermittelt. In diesen Units findet eine intensive, systematische Hinführung und Vorbereitung auf die Anforderungen der schriftlichen und mündlichen Abschlussprüfungen statt.

In den Units 11 – 12 (Special topics) liegt der Schwerpunkt auf interkulturellen Aspekten. Neben verschiedenen Textsorten zu diesem Thema werden hier auch ganz konkrete Verhaltensregeln vorgestellt und diskutiert.

In den Units 13 – 14 (Special topics) werden Situationen im beruflichen Alltag behandelt, z.B. Bewerbung und Lebenslauf verfassen, telefonieren, Buchungen und Reservierungen vornehmen.

In den Units 15 – 24 werden Texte aus unterschiedlichen Themenbereichen behandelt und mit prüfungstypischen Aufgabenstellungen bearbeitet, die sich am Gemeinsamen Europäischen Referenzrahmen für Sprachen und an neuen Prüfungsordnungen orientieren.

Die Aufgaben sind in allen Units nach den Kompetenzbereichen des Europäischen Referenzrahmens gekennzeichnet:
R = Rezeption **I** = Interaktion **P** = Produktion **M** = Mediation

Im Anhang befinden sich umfassende Lernhilfen (Skills Files), eine ausführliche Grammatikübersicht, das kapitelbegleitende Vokabular mit Phonetik, die alphabetische Wortschatzliste sowie eine Liste der unregelmäßigen Verben.

Auf dem Klappumschlag sind darüber hinaus Redemittel für die schriftliche und mündliche Arbeit im Unterricht aufgelistet.

Freeway – Allgemeine Ausgabe – zeichnet sich durch folgende Merkmale aus:

■ Übersichtlichkeit und schnelle Orientierung durch das Doppelseitenprinzip.
Die Units 1 bis 10 bestehen aus:

Starter	**Einführung in das Thema**	Auf der mit Fotos und Zeichnungen illustrierten Motivationsseite findet eine erste Annäherung an das Thema der Unit statt. Bereits vorhandene Kenntnisse werden durch einfache Übungen aktiviert.
Teil A (Doppelseite)	**Textarbeit**	Anhand von Texten, Statistiken und Übersichten wird in das Thema der Unit eingeführt. Die Grammatikschwerpunkte werden kontextualisiert vorgestellt und imitativ bzw. reproduktiv geübt.
Teil B (Doppelseite)	**Grammatik**	Im B-Teil wird der Grammatikschwerpunkt der Unit trainiert. Die Übungen sind inhaltlich an die Unit angelehnt, das Niveau des Wortschatzes wird hier aber bewusst niedrig angesetzt, um die Grammatik flexibel einsetzen zu können.
Teil C (Doppelseite)	**Textarbeit**	Der C-Teil dient der Vertiefung des Unitthemas, wobei in der Regel ein längerer Text zu bearbeiten ist. In den Units 6 – 10 finden sich hier Aufgaben zur Prüfungsvorbereitung.
Teil D	**Hörverständnis und Sprachproduktion**	Der D-Teil schließt jede Unit mit Übungen zum Hörverständnis ab. Dazu werden hier neben produktiven handlungsorientierten Aufgaben Sprechanlässe zur Verbesserung der Kommunikationsfähigkeit geschaffen.

■ Flexible Handhabung
Um den Vorkenntnissen und Interessen der Lernenden gerecht zu werden und um die Ziele der jeweiligen Schulform optimal zu erreichen, soll der Unterrichtende die Möglichkeit haben, flexibel mit Freeway zu planen.

Dabei sind z.B. folgende Varianten denkbar:
- Bei leistungsstarken Klassen könnte es sich z. B. bei den ersten fünf Units anbieten, nur die C-Teile zu bearbeiten (Vertiefung des Themas).
- Ebenfalls wäre es möglich, jeweils nur den A- und B-Teil (Einführung des Themas und Grammatikschwerpunkt) zu behandeln.
- Soll verstärkt die Grammatik geübt werden, dann ist auch ein Schnelldurchgang möglich, indem lediglich der B-Teil durchgenommen wird.
- Die Hörverständnisübungen und weiteren Aufgaben im D-Teil können natürlich mit allen Varianten kombiniert werden.

Klare grammatikalische Strukturierung
Der Schwerpunkt der Grammatik liegt in den ersten Units auf der Wiederholung der wichtigsten Tenses, in den folgenden Units in den Bereichen **Conditional, Passive voice, Reported speech, Participle, Gerund und Infinitive.**
Um die Übersichtlichkeit im Lehrbuch zu erhöhen, wurden Übungen zu Nebenstrukturen in das Workbook ausgelagert. Eine benutzerfreundliche Grammatikübersicht (Grammar Files) mit Erklärungen befindet sich im Anhang.

Intensive Textarbeit und Prüfungsvorbereitung
Auf der Grundlage eines längeren Textes dient bereits der C-Teil in den Units 6 - 10 der Vorbereitung auf die Aufgaben der Prüfung mit gezielter Hinführung zu Textverständnis und Textproduktion.

In den Units 15 - 24 (Exam preparation) liegt der Schwerpunkt auf der Bearbeitung von unterschiedlichen - zum Teil auch längeren - Texten mit Hilfe eines prüfungsrelevanten Aufgabenapparats. Zahlreiche Zusatzübungen zu den Texten (mit Fragen zum Text, zum Wortschatz usw.) finden sich im Arbeitsbuch.

Skills Files
Lerntipps und Hinweise zu **Reading, Writing, Speaking, Describing** und **Project Work** sind in den **Skills Files.** Verweise auf die **Skills Files** befinden sich in den Units (→ **SF**).

Weitere Merkmale
- Ein inhaltlicher und kommunikativer Schwerpunkt wurde auf die Bedeutung und Verwendung der englischen Sprache in einem zusammenwachsenden Europa gelegt.
- Unter Berücksichtigung des handlungsorientierten Ansatzes wurde versucht, die Lernenden anhand unterschiedlicher Situationen an adäquate Kommunikationsformen heranzuführen.
- Fester Bestandteil der ersten 10 Units sowie der Units 15 - 20 (B-Teil) ist die Schulung des Hörverständnisses unter Verwendung einer Vielzahl von Sprechern mit verschiedenen Akzenten (auch non-native speakers). Ausgewählte Texte des Schülerbuches wurden ebenfalls von verschiedenen Sprechern gelesen und befinden sich ebenfalls auf der CD.

Das Lehrwerk Freeway - Allgemeine Ausgabe umfasst das vorliegende Lehrbuch, ein Arbeitsbuch und ein Lehrerpaket, bestehend aus einem Lehrerband mit didaktisch-methodischen Hinweisen und Hintergrundinformationen, dem Schlüssel zu allen Aufgaben des Schülerbuches und den Texten aller Hörverständnisaufgaben. Zwei Audio-CDs mit den Hörverständnistexten und ausgewählten Lesetexten des Lehrbuches begleiten das Lehrwerk. Eine Besonderheit des Lehrerpakets ist die ebenfalls darin enthaltene Service-CD-ROM mit Texten und Vokabeln aus dem Schülerbuch sowie einem Übungsgenerator.

Wir wünschen Ihnen viel Spaß und Erfolg bei der Arbeit mit Freeway - Allgemeine Ausgabe.

Autorin/Autoren und Redaktion

Audio-CDs

Listening comprehension

Reading texts

→ SF 1 Skills File 1 (Appendix)

World Time Zones
Date Line
Deviation in hours
-12 -11 -10 -9 -8 -7 -6 -5 -4 -3 -2 -1 0 +1 +2 +3 +4 +5 +6 +7 +8 +9 +10 +11 +12
World languages
English
Spanish
Portuguese
French
Arabic
Chinese
Countries with economic agreements:
NAFTA
G8
EU
Borders of the economic areas
ARCTIC OCEAN
GREENLAND (Kalaallit Nunaat) DENMARK
GREENLAND SEA
BARENTS SEA
NORWEGIAN SEA
BEAUFORT SEA
Baffin Bay
Davis Strait
Denmark Strait
Hudson Bay
LABRADOR SEA
Gulf of Alaska
ALASKA UNITED STATES OF AMERICA
CANADA
UNITED STATES OF AMERICA
MEXICO
Gulf of Mexico
CUBA
NORTH PACIFIC OCEAN
NORTH ATLANTIC OCEAN
SOUTH ATLANTIC OCEAN
SOUTH PACIFIC OCEAN
INDIAN OCEAN
VENEZUELA
COLOMBIA
ECUADOR
PERU
BRAZIL
BOLIVIA
PARAGUAY
CHILE
ARGENTINA
URUGUAY
ICELAND
NORWAY
SWEDEN
FINLAND
UNITED KINGDOM
IRELAND
FRANCE
SPAIN
PORTUGAL
GERMANY
POLAND
ITALY
UKRAINE
BELARUS
TURKEY
RUSSIAN FEDERATION
KAZAKHSTAN
MONGOLIA
CHINA
JAPAN
N. KOREA
S. KOREA
IRAN
IRAQ
SYRIA
SAUDI ARABIA
AFGHANISTAN
PAKISTAN
INDIA
MEDITERRANEAN SEA
MOROCCO
ALGERIA
LIBYA
EGYPT
MAURITANIA
MALI
NIGER
CHAD
SUDAN
NIGERIA
ETHIOPIA
SOMALIA
KENYA
TANZANIA
DEM. REP. OF CONGO
ANGOLA
ZAMBIA
NAMIBIA
BOTSWANA
MOZAMBIQUE
MADAGASCAR
SOUTH AFRICA
INDONESIA
PAPUA NEW GUINEA
PHILIPPINES
MALAYSIA
AUSTRALIA
NEW ZEALAND
TASMAN SEA
SEA OF OKHOTSK
BERING

Training for work

LEEDS COMPREHENSIVE SCHOOL

GCSE
General Certificate of Secondary Education

LEEDS COLLEGE

Engineering studies

Social studies

Business studies

BTEC
National Diploma

Public Service

Company

University

a Look at the photos and answer these questions.

1. What do the photos show?
2. What can the students from Leeds Comprehensive School do at Leeds College?
3. What possibilities do they have after college?

b Why are you at college?

1 Come to Leeds College

LEEDS COLLEGE

INFORMATION | PROGRAMME OF STUDY | LINKS | COURSES | NEWS | OPEN DAYS

EMAIL: INFO@LEEDS-COLLEGE.AC.UK
WWW.LEEDS-COLLEGE.AC.UK

Our college is one of the largest in the country. We employ over 400 teachers, both full-time and part-time. Our students are of all ages. About half of them are under 20, but we even have one who is 86 years old. The college runs over 200 courses at all levels. Most of the study programmes take place between nine and five, but we also offer evening and part-time courses.

BTEC National Diploma courses
These full-time courses lead to higher job qualifications. Successful students can also go on to university. Changes are taking place in Europe. On these courses students can learn a lot about Europe and study a European language. This is a great start for a career in a modern Europe!

Costs:
The courses are free for students under 18. Students who live over 3 miles from the college receive a free bus pass. Students over 18 can get a grant to cover part of their costs for books and equipment.
For further details see our homepage, send us an e-mail or write to:

Leeds College
Shaftesbury Road
Leeds LS2 45

Tel. 0532/449536

2-YEAR-PROGRAMME OF STUDY

Engineering Studies (Electrical, Mechanical or Civil)
- Mathematics
- General science
- Electrical/mechanical principles
- Computer studies
- English
- European language (French or German)
- Engineering applications of computer systems (CAD & CAM)
- Industrial control and instrumentation

Business Studies
- Mathematics
- General office skills
- Accountancy
- Computer studies
- English
- European language (French or German)
- Statistics
- European business and finance

Social Studies
- Sociology
- Psychology
- Principles of education
- Health care
- Statistics
- Computer studies
- European language
- English

a Answer the questions below. Start your answers with expressions like: R

"In the text/brochure it says …"
"In the second paragraph you can find out …"
"In line … of the first/second … paragraph you can read …"

1. How do you know that Leeds College is very big?
2. How old are the students there?
3. When do most of the students have their lessons?
4. Why do the students learn a European language?
5. What can you get if you live far from the college?
6. How can people get in contact with Leeds College?

b Talk about your college. Think about: P
size of college / teachers / students / courses / your course / costs

c Make a small brochure or a homepage for your college. P

2 Two students meet on the campus

Terry: Hello, Sally. Long time no see. What are you doing here?

Sally: I'm doing a BTEC course.

Terry: And how are you finding it here?

Sally: Right now my big problem is mathematics. There are so many new formulas to learn. And then of course, there's the money.

Terry: I see what you mean. I've got a part-time job with my old firm. I work in the packing department for four hours in the evening. It's hard work, but it helps me pay for my flat.

Sally: Oh, it's easier for me. I still live with my parents. But I've got a weekend job in a restaurant. That gives me a little more pocket money.

Terry: Which department are you in, by the way?

Sally: I'm in the Electrical Engineering department. First year. And you?

Terry: I'm in the Business Studies department. In my second year. We're studying accountancy, statistics and subjects like that at the moment.

Sally: What foreign language are you learning on your course?

Terry: This year I'm doing German for beginners – and that's my problem. There are so many new words to learn. Guten Tag. Ich … heiße Terry. … möchtest du …

Sally: Wow, look at the time. It's eleven o'clock. I've got a science lesson at five past. I have to rush. See you.

Terry: Bye. … eine Tasse Kaffee?

a Copy the information sheet about the two students and complete it. R

	Student 1 (female)	Student 2 (male)
Name?		Terry
Course?		
Department?		
Home?		
Problems?		

b Now write about the two students using the information above. P

c Use the expressions in the dialogue and make a new dialogue between these two students. P

Mike
- 1st year BTEC course in mechanical engineering
- bedsitter (one-room apartment)
- weekend job as a tennis coach
- problems with French/money

Linda
- 2nd year BTEC course in social studies
- house with three other students
- evening job as a waitress at a local club
- problems with statistics/money

1 Jill is a student

This is what she says about herself:
"I'm eighteen and I go to Leeds College. I don't have any brothers or sisters. I live in Leeds with my parents. My parents give me £100 a month. I spend most of the money on books and things for college. In the evening I work at Pete's Pizza Palace. I need the money, because I buy quite a lot of clothes. In my free time I don't go out very often.
I usually go to college by bus. Sometimes a friend takes me there in his car. At college we have a lot of subjects. I'm in the second year. I like English and mathematics best, but I don't like French. We don't have German. Very often I do my homework with a friend. Normally we both work hard for our course. It leads to higher qualifications and I hope to find a good job afterwards."

a Now talk about Jill.

Example:
"Jill is eighteen and she goes to Leeds College. She doesn't have any brothers or sisters. She …"

b Do you remember? Ask your partner(s) questions about Jill. Do not look at the text on the left.

Example:
Where – Jill – live?
Where does Jill live?
Jill lives in Leeds.
Who – live – with her parents?
Who lives with her parents?
Jill lives with her parents.

1. Where – Jill – go to college?
2. Who – give – her – £100 a month?
3. What – Jill – do – in the evening?
4. Why – she – work?
5. How – she – get – to college?
6. Which subjects – Jill – like best?
7. Who – do – homework – together with her?
8. How – they – work – for their course?
9. Which subject – they – not – have?
10. What – Jill – hope to find?

2 In the cafeteria

a Guess what the students in the picture are doing.

Example:
Helen's writing.

b What do you think? Ask your friends questions about the students in the cafeteria. What are they doing exactly?

Use: What • How many • Where
What … about • Who … to

Example:
What is Helen writing?
I think she's writing a letter.

3 People at Leeds College

John is a student and Lucy is a secretary.

- usually
- normally
- on weekdays
- from Monday to Friday
- every day
- often

- do homework
- organize meetings
- help visitors
- listen to teachers
- take a lot of calls
- do tests

today …
right now …
at the moment …

■ Talk about John and Lucy.

1. What do they normally do?
2. What are they doing at the moment?

4 On the phone

■ Fill in the right tense.

Sally: Hi, Tom. Are you busy at the moment?
Tom: Yes, sure. Why?
Sally: What **1. (do)** you?
Tom: I **2. (do)** my homework, of course.
Sally: What again? You never **3. (do)** anything else, Tom. Why you **4. (stay)** home all the time?
Tom: You know that I **5. (prepare)** for my exams and right now I **6. (try)** to do my mathematics. Our teacher **7. (give)** us more than 20 exercises every day.
Sally: And what about me? I **8. (wait)** for your call every evening.
Tom: I'm really sorry about that. Why you **9. (not go out)** with Susan sometimes?
Sally: I **10. (not see)** her very often, because she **11. (study)** for a different exam this year. And Anna **12. (live)** in France at the moment, because she **13. (work)** on her French project. I'm so unhappy …

5 Money problems

a The quotations in the bubbles on the right are exceptions to the rule. Explain why.

b Now fill in the right tense in the following sentences.
1. "As you **(know)** life **(get)** so expensive these days, especially for poor students like us. **2.** I always **(tell)** Jack – why **(go)** you shopping at the corner shop? **3.** Things **(cost)** much less at the supermarket near the college. **4.** It **(annoy)** me because I **(get)** the feeling that he **(not listen)** to what I **(say)**. It's terrible. **5.** We … always **(have)** arguments about things like that. **6.** Sometimes I **(think)** he just **(be)** lazy. **7.** On the other hand there are other things where Jack **(think)** we can save. **8.** We **(own)** a little car for example. **9.** Jack **(say)** we **(not need)** it. **10.** He **(be)** quite right. **11.** And so we **(think)** of selling it. **12.** We **(have)** to do something. **13.** Money **(not grow)** on trees."

"Is your daughter having a good time at college?"

"You know, students are always complaining about money but every third student owns a car."

"Nowadays a lot of students are thinking seriously about their future and things are getting more expensive. And as you know, money makes the world go round."

1 Leeds and Dortmund: An exchange programme

City plays host to German guests

by Marigold Prince

"For these young people it's quite an adventure" says head of department Willi Schmitz from the Berufsbildende Schulen in Dortmund, Germany. His group of 16 students are just getting to know their English host partners at a reception at Leeds College.

"For many of them it's their first visit to Britain and their first chance to put their English to the test." Indeed, the group of 18–20 year-old students don't know very much about each other before their first meeting. Usually students in both groups receive a passport photo of their partners and a short introduction form before the exchange. This form gives them information about their partner's leisure interests, plans for the future and their favourite food and music.

"I hope mine's not a smoker who's only into heavy metal!" remarks one young Leeds student to his friend as he passes me with some coffee and biscuits for the visitors.

As well as the new situation with their partners and the language, for one week the group has to cope with the different way of life in an English family. "But," adds Willi Schmitz, who is on his fourth visit to Leeds, "after they get to know each other on the first evening, they usually all get on like a house on fire."

This week their programme includes visits to a number of local firms, trips out to the Yorkshire Moors and the Lake District (including a stop at Borwick Hall on the way) as well as a joint project in the College workshops. "This year both groups are working together on a business studies and technology project on local environment problems," says Peter Stewart, Leeds College's exchange organizer.

"We find that working together on a project is certainly one of the best ways of making really good friends. Many of the students from past exchanges still visit their partners or go on holiday with them and every year there are always one or two students who are especially unhappy when it is time to leave," Peter tells me.

As I leave the staffroom at the College, the German students are collecting their luggage and leaving for the first night with their partners and their families. Already it seems to me that some of their earlier ideas about Britain and the British are changing. I hear a couple of comments from some of the German students. One of them says, if I understand his German correctly, "They seem really friendly." "Yes, and it's not raining, either," replies his friend and smiles. Maybe travel really can broaden the mind!

435 words

a Look at the picture which goes with the newspaper article. Who do you think the people are?

R **b** Read the article and check if you are right.

c Use information from the text and make a list of four **P**

1. people the students meet.
2. things the students may worry about before they leave Germany.
3. things on the programme in Leeds this week.

R **d** Which of these statements are true and which are false? Explain why. Use information from the text.

1. The first exchange visit to another country is always an exciting experience.
2. Exchange students are well-informed about each other before the visit.
3. One of the Leeds students wants a smoker as a partner.
4. After one meeting the exchange students generally like each other very much.
5. There are visits to local sights on the week's programme.
6. Students who work together get to know each other well.
7. There are two projects this year, one for the business students and one for the technology students.
8. Students who take part in these exchanges do not often see each other again.

R **e** Which three statements in exercise "d" represent the main ideas in the text?

R **f** In these sentences from the text, who or what do the words in bold type refer to?

1. … **it's** quite an adventure … (line 1)
2. For many of **them** it's their first visit … (line 7)
3. … **he** passes me with some coffee … (line 25)
4. … who is on **his** fourth visit to Leeds … (line 30)
5. As **I** leave the staffroom … (line 56)
6. … **they** seem really friendly … (lines 63/64)

g Find the opposites of these words in the text. R

guest; work; same; last; arrive; alone

h Here are some words from the text and their dictionary definitions. R
Find the matching pairs.

1. host	a) something people do together
2. to get to know	b) when things or people change places
3. leisure	c) to make something wider or bigger
4. to cope with	d) person who has guests
5. exchange	e) to meet someone for the first time
6. joint (project)	f) something difficult or dangerous
7. to broaden	g) to manage a situation well
8. adventure	h) free time

i Now answer these questions. R

1. Why is an exchange visit such an adventure?
2. Why is Willi Schmitz sure that his students can have a good time in Leeds?
3. Why is it important for the students to work together on a joint project?
4. Why do students from the past exchanges still visit each other?
5. Why does one student say "Yes, and it's not raining, either"?

j Translate lines 56 to 66. → SF 10 M

2 A personal letter

You are taking part in an exchange scheme that your college is organizing.
You already have some information about your partner and now you want to write to him or her to introduce yourself.

P Write an informal letter in English. → SF 12
Begin with 'Dear …' and end with 'Yours …' or 'with love from …'

Include these points:

- yourself and your family
- your interests
- what you like/dislike
- your personality and what you look like
- your town
- the weather in your area

Use these expressions:

I am … /in my family we usually …
I am interested in …
I like/love/can't stand …
Some people say I am …/I look …
There's …/we've got … in our town
it's sometimes …/often … here

1 An exchange visit

a Listen and then answer these questions. R

1. Who are the speakers?
2. Who is talking about what? Look at the pictures for help.

b Listen again and correct these statements. R

1. The exchange students are working on the environment project at the end of the week.
2. John thinks that visits to local firms and factories are boring.
3. Klaus thinks you learn more about the British as a tourist.
4. John's mother is not very enthusiastic about the exchange.
5. John believes that the Germans are all serious and hard-working.
6. Klaus finds living in an English family strange.

c Complete the sentences with the phrases in the box below.

1. … it's really great that we have the chance to see places like that.
2. …, that's why these exchanges are such a good idea.
3. Oh yes, … with you.
4. … that tourists learn half as much.
5. They don't have enough contact with the locals, …
6. They still think, …, that all the Germans are serious.
7. …, isn't it a bit strange for you, Klaus, living in a foreign family?
8. …, you're all so friendly to me.

I suppose • I think • I believe • I'm sure • in my view • I agree • in my opinion • I'm quite certain • I don't think • for example • for instance • I mean

2 Managing situations: What would you say? → SF 10 M

1. Sie sind in England und lernen jemanden auf einer Party kennen.
2. Sie sagen, dass Sie Austauschschüler(in) aus … sind.
3. Sie fragen Ihren Gesprächsparter/Ihre Gespächspartnerin, ob er/sie zur Schule geht oder arbeitet.
4. Sie erkundigen sich nach seiner/ihrer Lieblingssportart.
5. Sie stellen Ihrem/Ihrer neuen Bekannten einen Mitschüler/eine Mitschülerin vor.

Just the job

Are people always asking you what you want to be when you leave school? Try our special chart to help you discover which is the job for you!

■ Work out the best job for your partner. He/she should answer 'yes' or 'no' to the questions. It's more interesting if your partner closes his/her book. When you reach one of the symbols, look at the list below for a suitable job.

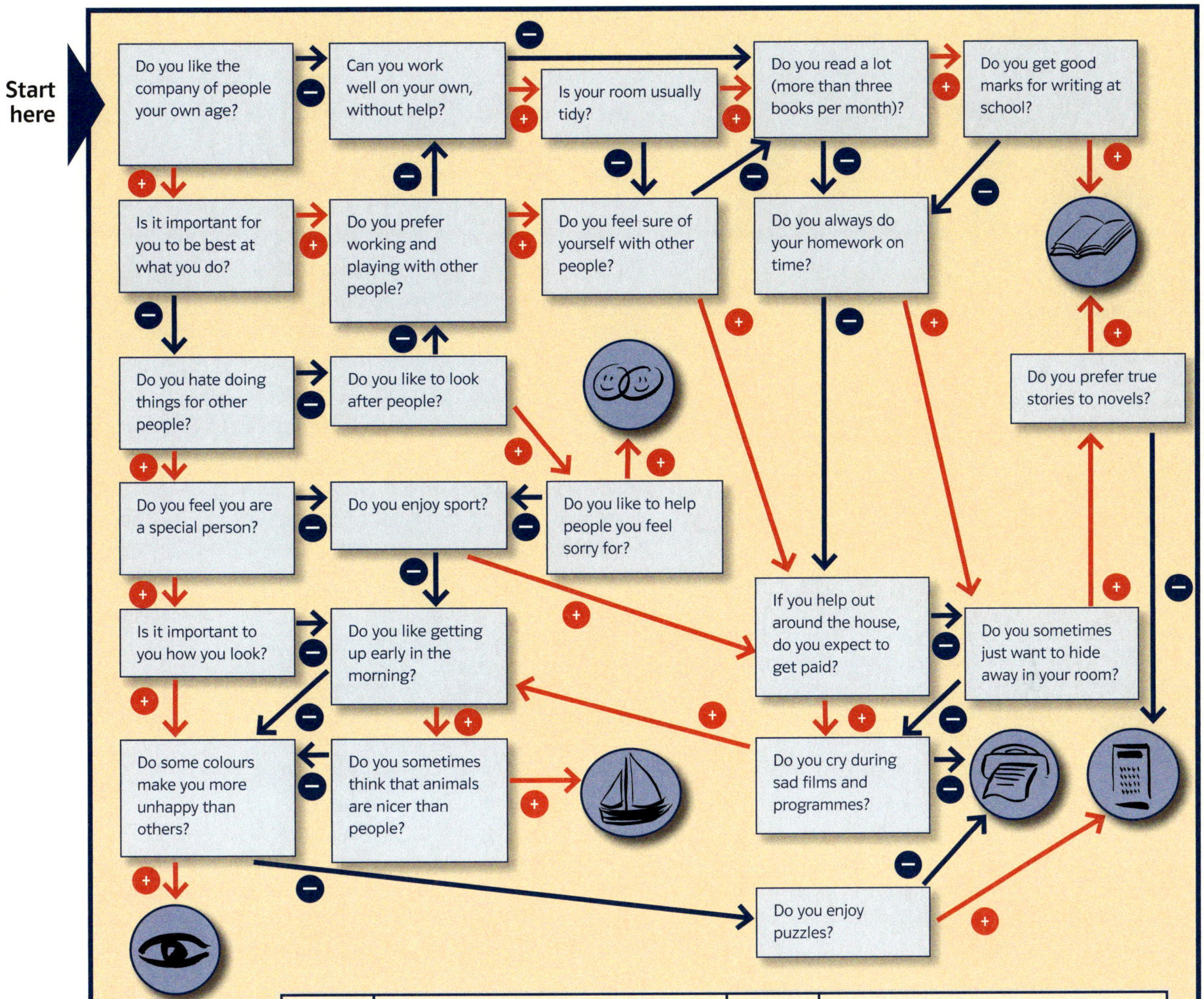

	**You're active and practical.** The job for you: something active like a sportsman/woman, police officer or mechanic.		**You're practical and creative.** The job for you: something like an office worker, tourist guide, salesman/woman or designer.
	**You're creative and like people and talking.** The job for you: something creative like a singer, dancer, journalist or hairdresser.		**You're clever and independent.** The job for you: one which has something to do with books like a language teacher or writer.
	**You're helpful and friendly.** The job for you: one where you can help people like a nurse, flight attendant or receptionist.		**You're logical and like to learn.** The job for you: one where you must solve problems and which is well-paid like a computer specialist or advertising manager.

1 "Computers Unlimited"

P **a** Look at the photo and describe what you can see. → **SF 19**

The American dream is alive and living in North London. John Sanson started his own business, Sanson Computers, 6 years ago in a small flat above his parents' butcher's shop. However, in the last few years his sales have increased so enormously that his company now has a turnover of £13.5m. At 28 Mr Sanson has already become a very rich man. Only 2 weeks before he opened his computer shop, he graduated from Nottingham University with a degree in Electrical and Computer Engineering. "A lot of people in my family were self-employed and I wanted to do the same. I have always been interested in computers and while I was at university I saw the possibilities of a growing market in personal computers. So, when I left university, I decided to take the risk," John explains. However, there were not a lot of risks for him; he lived with his parents and was used to life on a student's grant. Moreover, he only had a staff of one, so he didn't have a high wage bill. Sanson's firm is now called 'Computers Unlimited'. He employs more than 80 people who sell and service PC software and hardware. "Nowadays you can go out and buy a computer like a TV or video recorder," says Sanson. As a result his business has expanded enormously. In his first year he made a profit of only £11,000. In the meantime his profits have reached £2.3m a year. Naturally enough, he has moved from his small shop in North London and has now rented a large store in a business park just outside London. He sums up his success story like this: "Of course it has been hard work. But the secret is that I still enjoy the work and my boss likes me, too."

302 words

(© Roger Trapp, The Independent, London, adapted)

R **b** Answer the following questions.

1. Where did John Sanson open his first shop?
2. What did John study at university?
3. Why did he decide to start his own business?
4. Why didn't John have a lot of risks at first?
5. How has his business developed?
6. Where has he moved to?
7. Why, do you think, has he moved?

R **c** The following words are in the text. Match them with their definitions.

d Find words or expressions in the text which have the same meaning as the following. **R**

1. company 2. little 3. grown
4. left university with a degree 5. workers
6. computer equipment 7. up to this time
8. employer

e Say whether the words you have found are nouns, adjectives, adverbs or verbs.

f Use the words to talk about John Sanson. **P**

1. to employ (line 22)	A. the technology of electricity (n.)
2. electrical engineering (line 10/11)	B. to give work to. They … 150 workers in that factory (vt)
3. self-employed (line 12)	C. to a great extent (adv.)
4. enormously (line 26)	D. (good fortune) to have great … in life (n.)
5. to increase (line 5)	E. to make or become bigger in size or number (vt)
6. success (line 31)	F. having your own firm or company (adj.)

2 A radio interview

"Young people in Southampton" is a popular programme on Radio Solent, the local radio station. Des Jacky is talking to Roberta Crumb.

DJ: Welcome to "Young people in Southampton". Today we have Roberta Crumb with us in the studio. Roberta, you are 24 and you ran a small video production service here in Southampton. Now could you tell our listeners a little about what has happened to you and your business?

RC: Well, it all started when I saw an advert in the local paper. At that time I was employed at a large electronics firm near here.

DJ: And what was the advert for?

RC: "Starting your own business." The local bank offered to give advice to young people who wanted to start up their own business. At first I didn't take it all that seriously. But the more I thought about it, the more interested I became. I talked to my parents and my boyfriend about it. They said I should do what I really wanted to do, which was to start up a small video production service. So, after some time I went to the bank and had a talk with the bank manager. He explained all the advantages and also the risks of being self-employed.

DJ: What happened then?

RC: Well, after a lot of thought I made up my mind to start up the business. We recorded customers' weddings, birthday parties – anything like that, but I didn't earn as much as in my old job.

DJ: And what happened then?

RC: Well, I still had to pay back a lot of money to the bank for the video equipment, and advertising cost me a lot, too. I couldn't live on the income from my work. So I decided to close down the business. I think my main mistake was that I didn't watch the market development carefully enough. People nowadays have their own video cameras and don't need a professional service.

DJ: And where are you working now, Roberta?

RC: I've got a job in a photographer's shop. The pay's not bad, but the problem is I have to pay back the money to the bank. I haven't had a holiday for 2 years now. I just can't afford one.

R **a** True or false? Say why or why not.

1. Roberta worked in a photographer's shop.
2. The local bank put an advert in the newspaper.
3. Roberta earned enough money from her business.
4. The bank manager gave her some advice on starting her own business.

R **b** Complete these sentences. Choose from the nouns and verbs in the boxes. Use the same tenses as in the dialogue.

1. Roberta … a video production …
2. She … an … in the local newspaper.
3. She … a talk with the …
4. Roberta … up the business.
5. She (not) … the market … closely enough.
6. Roberta (not) … enough customers.
7. She … down her …
8. She … in a … shop.

have (2) • see • run • start • watch • work • close

advert • development • service • business • bank manager • photographer's

P **c** Join the sentences with the linking words in the box.

Example:

1. In the interview we hear that Roberta ran a video production service.
2. First she …

In the interview we hear that …
Then … After that …
However … As a result …
So … Now … First …

1 Softright Systems

Jane Topper moved to Glasgow seven years ago. There she met Harry Coe who didn't like his old job in an import company. As they were both interested in computers they **1. (decide)** to start a small software company. First they only **2. (write)** computer programs for Harry's old import company, but soon they **3. (see)** that a lot of other companies also **4. (need)** special software. So they **5. (hire)** some young programmers who **6. (work)** in small teams and **7. (develop)** a lot of new ideas.

The company **8. (be)** profitable from the start. Although other companies **9. (have)** their problems Softright **10. (not lose)** money. In their third year in business they **11. (make)** a profit of £150,000. Then they **12. (try)** some advertising and it **13. (work)**. That is why they **14. (spend)** more on marketing in the following years. At the beginning Jane and Harry only **15. (have)** two small rooms for their work. After some time they **16. (rent)** offices in several different buildings. But last year they **17. (buy)** a warehouse in central Glasgow. The move **18. (not be)** easy and it **19. (take)** more than two months. "We **20. (lose)** quite a lot of money when we **21. (move)**," Jane said. "But we **22. (not want)** to work in different places any more. That **23. (be)** a good decision!"

a Complete the sentences in the past tense.

b Find the missing parts in the following statements about Jane and Harry's company. Ask a classmate about it. He/she should give you an answer with the help of the information in the box below.

Example:

1. When did Harry's parents move to Glasgow? They moved to Glasgow in 1976.
2. Where …

1. Harry's parents moved to Glasgow in …
2. After college in Glasgow Harry worked at …
3. The company imported …
4. He enjoyed his … very much.
5. But one day Harry left because …
6. He met … on a computer course.
7. She told him …
8. Of course, they needed … for their new company.

- some money
- that she wanted to start her own business, too
- he wanted to be his own boss
- his partner
- 1976
- his father's company
- electronic parts
- work

2 Times have changed

Describe what has happened to the following two companies in the last few years.

Example:

Evans & Sons have been in Sheffield since 2000. Linda Graves has …

	be in Sheffield	sell	make a loss	expand	buy a company abroad
Evans & Sons	since 2000	120 million DVDs so far	never	three times	already
Linda Graves	for five years	about 1,000 cars up to now	once	just	not yet

3 A job interview

Mary Wilson has applied for a job in the export department of Kwik-Fit. In an interview she talks about her life and career.

a Take her role. Use the verbs and the information below.

Example:
"I was born in Leeds in 1979. I went to school from …"

be
pass
get
live
go
study
travel
work

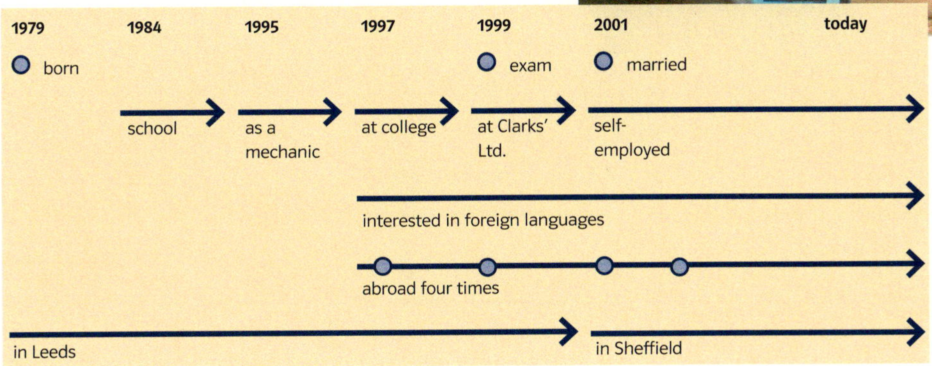

b Ask your classmates about:
date of birth / school / college / important exams / address / trips abroad / experience in English / experience at work / hobbies

c Complete the following report about Kwik-Fit. Put the verbs into the simple past or present perfect tense.

Example:
35 years ago Tom Farmer opened two garages in Edinburgh. In 1974 he …

Kwik-Fit

35 years ago Tom Farmer **(open)** two garages in Edinburgh. In 1974 he **(sell)** to G.A. Robinson. Shortly afterwards Robinson **(get)** into difficulties. Then Farmer **(buy)** the company again. Since then the company's name **(be)** Kwik-Fit. In the beginning most of the business **(be)** with private cars, but in 1987 Kwik-Fit **(begin)** a service for companies. In the last few years customers **(include)** ICI, Avis and British Petroleum. Up to now Kwik-Fit **(keep growing)**.

4 How long … ?

a Answer the following questions.

1. How long have you been working with this book? **for … weeks/since …**
2. How long have you known about this college? **for … years**
3. How long have you been learning English? **for … months/since …**
4. How often have you visited England? **once, twice, … times, never**

b Translate the following sentences. Watch out for the correct tenses. → SF 10

1. Peter arbeitet seit Januar bei Seafood Limited.
2. Er kennt den Manager seit einigen Jahren.
3. Sie planen seit drei Monaten ein neues Projekt.
4. Peters Frau hat sich 3-mal um eine Stelle beworben.
5. Sie wartet schon seit langem auf eine Antwort.
6. Sie sind schon lange nicht mehr zusammen im Urlaub gewesen.

1 Workers – past and present

R **a** In one sentence say what subject these texts have in common.

For the people who lived in the East End of London in the middle of the nineteenth century, life was hard. Often a whole family lived in one room and even 12-year-old children went out to work. Working days were long, often from five in the morning until nine at night. There were no holidays until 1871 when the government introduced the first Bank Holidays. At this time 77 % of the population of Britain belonged to the working class and so had to live in such conditions. Even to buy the most basic food could cost a man three days' pay. When a man lost his job, which could happen at the end of the week, the day or even the hour, or when he became sick, he had nothing except his family and friends to help him.

Europe's workers – underpaid but happy!

This is the result of a survey which a London firm has completed for the managers of a number of multinational companies who have employees all over Europe. The survey shows companies what they can expect from the workers in each country. More than 500,000 workers answered the questions about attitudes to pay, workmates and benefits.

The survey shows that in general Europe's workers enjoy what they do, get on well with their workmates and identify with the company they work for.

However, attitudes to pay are less positive. Although 57 % of workers in the Netherlands are satisfied with their wages, in Switzerland only 44 % of the workers are happy with their pay. In Britain the number is much lower, 35 %. Moreover, the Dutch also feel happiest about their working conditions and benefits. Workers there receive paid sick leave after two days at 70 % of their pay for one year.

The European Union has introduced laws about the maximum number of working hours. However, working hours within Europe still vary quite a lot. In Britain and Luxembourg the maximum is 48 hours, but in Denmark it is much lower, 37 hours.

There are also laws in the EU about the minimum amount of paid holiday a worker can have. The average in Britain is usually 23 days. However, German workers can expect a lot more than this.

Finally, on the question of job satisfaction, the survey's results are positive. Most of Europe's workers are happy with their workmates as well as with the job they do.

262 words

(The European, London, adapted)

Are you happy with your pay?

Are you satisfied with your job?

R **b** Which of these pictures show conditions of work which are mentioned in the texts? Name them.

1
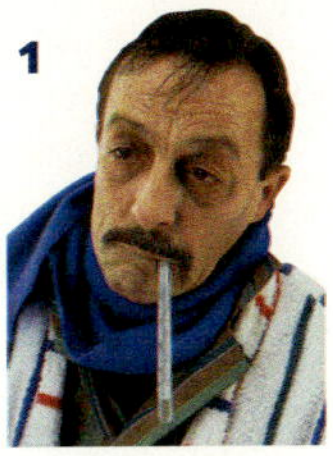
2

3

4

5

6

R **c** The following numbers appear in the texts. What information do they give?
five hundred thousand; twelve; two; thirty-five; five; twenty-three

R **d** Answer these questions.

1. When did 77% of the British population belong to the working class?
2. How many hours per day did people often work in the past?
3. What help did workers in the past receive when they became ill?
4. Why did the London firm carry out its survey?
5. The survey shows that most workers in Europe agree about some conditions of their work. Which ones?
6. Which two conditions of their work are the Dutch the happiest about?
7. Which conditions of their work do you think the British are unhappy about?
8. Job satisfaction depends on pay. Do you agree? Use the charts to help you with your answer.

P **e** How has the situation of workers in Europe changed in the last one hundred years?

M **f** Translate lines 1 – 8 (Europe's workers – underpaid but happy!). → SF 10

2 Your job?

P **a** Draw this list in your exercise book. Then list the jobs below under a suitable heading.

Health/Care	Office/Sales	Leisure/Tourism/Media	Technical jobs/Construction
1.	1.	1.	1.
2.	2.	2.	2.

nurse – engineer – doctor's assistant – insurance broker – accountant – car mechanic – welder – social worker – nursery school teacher – IT specialist – plumber – shop assistant – flight attendant – musician – care worker – carpenter – hairdresser – physiotherapist – bank clerk – secretary – office clerk – travel agent – fitness instructor – journalist – waiter – cook – electrician – builder

P **b** Work in small teams. Use the Internet to find at least three more jobs for each section. Compare your results and complete your lists in class. → SF 22

P **c** Which job would you (not) like to do? Why?

Job advertisements

A leading European engineering company requires responsible people to work in several departments:

TECHNICAL ASSISTANT
EXPORT ASSISTANT
CHILDCARE ASSISTANT

The successful candidates must be aged 21-30, have a BTEC National Diploma or similar qualifications and some work experience.
A good knowledge of English and German is essential in our international business.
Technical and export assistants should be willing to travel overseas. The childcare assistant should have experience in caring for children under 5.
The company offers interesting jobs with attractive salaries and excellent working conditions.

Applications with full CV to the personnel manager
Mr Len Linscott.

Always one step ahead!

IEP Industries
105-110 Oldham Road
Manchester M13 9PL
Tel: 0161-8342707
www.iepinfo.co.uk

R **a** Answer the following questions.

1. What is the firm looking for?
2. What qualifications must the applicants have?
3. What advantages can you find in the jobs?

P **b** Apply for one of the jobs above. Complete the missing information in the letter below. Use your own personal details, where possible.

name and address
of employer above

your address
date

Dear Sir/Madam/Mr .../Ms ...,

With reference to your advert in today's newspaper, I would like to apply for the vacancy of **(job)**. I am **(age)** and of **(...)** nationality. I have attended **(college in ...)** since **(date)**. In **(date)** I hope to pass my examination in **(department)**.
While I was at college/Before college I was employed at **(name of firm/institution)** from **(date)** to **(date)** as a **(job)**. In my job there I **(tasks)**. I am applying for the job as **(name of job)** in your company because **(reasons for applying)**. Enclosed please find my CV and copies of certificates.
I look forward to hearing from you soon.

Yours faithfully/Yours sincerely,
(name)

P **c** After you have written the letter of application, the next step is to write a curriculum vitae (CV). Make a list of all the information you need to include in your CV. Discuss your lists in class.

d Mr Linscott is interviewing two candidates for the job of Technical Assistant. Listen and find out: R

1. names of the candidates
2. their favourite subjects at school/college
3. their present jobs
4. why they have applied

e Decide who you would choose for the job. Give reasons. P

Young people in Europe

A survey in Europe: Do you feel European?

A survey released by the European Commission in 2003 shows how Europeans feel about this question. This is what they said:

Source: Standard Eurobarometer 2003

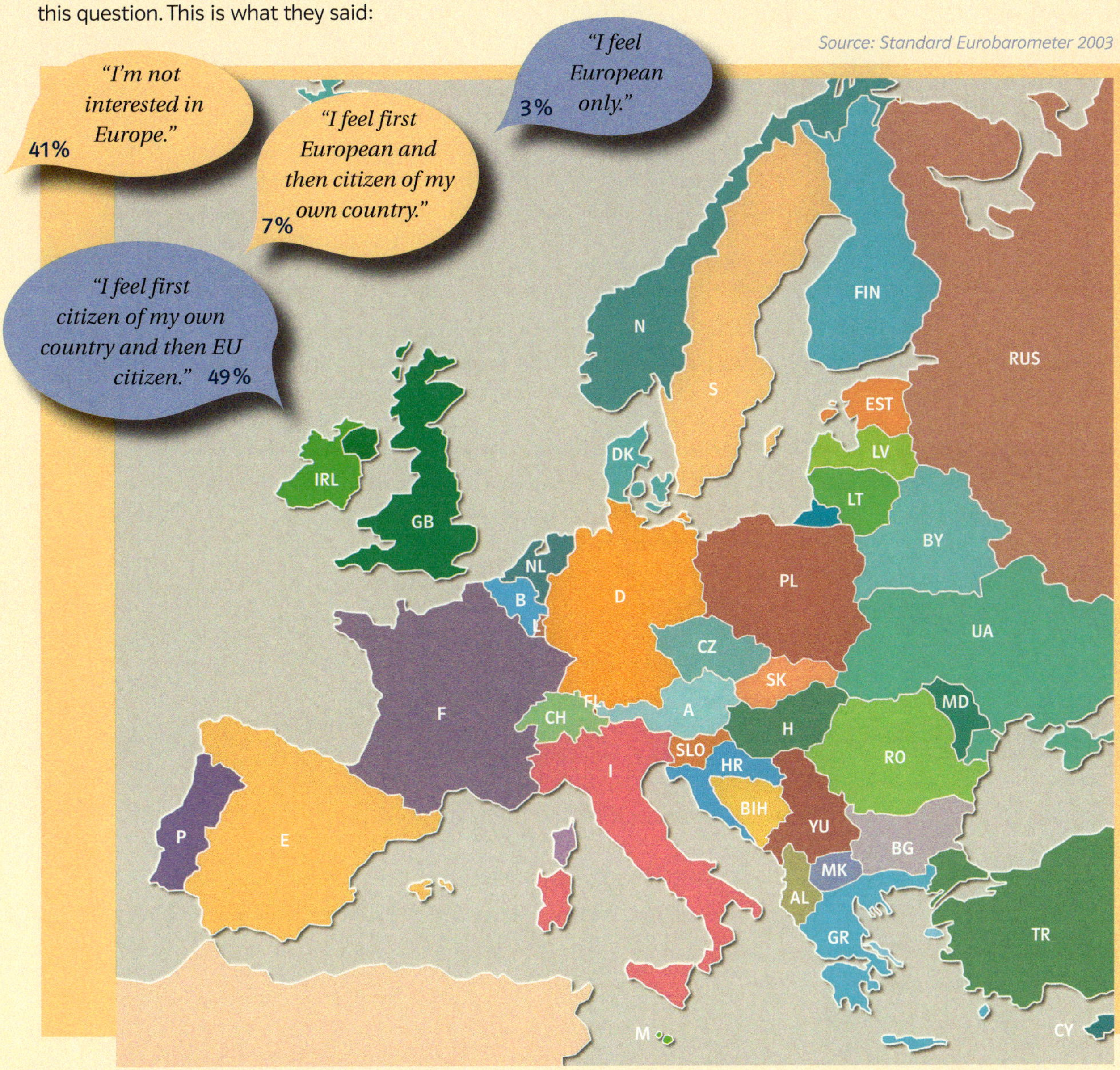

P **a** **Put the letters of these words in the right order to find the names of some European countries.**
canfre; dianler; laity; yemnarg; desnew; umigleb; cegeer; lopdna

R **b** **Answer the following questions.**
1. Which letters on the map show which countries?
2. What is the capital city of each country?
3. What are the most important languages in Europe?
4. Which of the countries on the map are now in the European Union?

c **Do you feel European? Look at the statistics above and make a similar survey for your class. Compare the results.** P

d **Find reasons for the differences.** P

1 Studying in Europe

Nowadays it's becoming more and more important to spend some time abroad, so if you want to study or train for a job, why not do so in another European country?

"When I arrived in Warsaw they had already organized everything for me," says 20-year-old Sven Berglund from Stockholm in Sweden. "A friend was waiting for me at the station and he took me to the student hostel where I stayed for the first three months. Now I share a beautiful flat with two other students. It's much more comfortable and costs much less than I would have to pay at home. Living in Poland is quite cheap, you know.
At the university in Warsaw, where I'm studying social studies, my sociology lectures are in English and German so I don't have to speak Polish. What I found a little strange in the beginning was that if you want to talk to your professor or practise discussions in seminars you have to wear a suit. I hadn't brought a suit with me so I had to buy one. But normally I wear jeans, of course."

"Don't take everything too seriously," says Brian Auger from Bristol. "That's what I first learned when I started my engineering studies in France. Most of the lectures are in French but I soon got used to that and, of course, it helped a lot that I had learned the language at school for many years.
I spent the first three weeks at a youth hostel here in Lyon because it wasn't easy to find a place to live. Then I got a couple of addresses from a students' organisation and was able to rent a small room. Also I didn't expect such a lot of bureaucracy in France – you have to fill in a lot of forms if you want to live here, and you need several kinds of insurance – and that's not cheap. But if you insure with a students' insurance company you get discounts in fast food restaurants and supermarkets. Isn't that great?"

"Studying in Great Britain can be expensive so money is the most important thing if you want to study here," says Esther Rosenkranz from Frankfurt. "First I applied for a place at Oxford University, but I didn't pass the entrance exam. Now I'm at the London School of Economics and I have to pay quite a lot of money a year, but on the other hand you study in small groups right from the start. That means that your professor knows you by name, which wasn't possible in Germany because there are often hundreds of students in the seminars there.
When I applied for a place in business studies I had to take a language test in English first, but I didn't find that too difficult. I've always enjoyed English. I'm really happy here and feel quite at home in the student hostel where I'm living. Luckily it's not too expensive and also it's very international, because quite a lot of the places are reserved for students from abroad, which I find exciting."

R **a** **Find at least 3 examples of each of the following in the text:**

1. countries
2. cities
3. subjects
4. languages
5. places to live

R **b** **Which of the three students is it?**

This student

1. found it very difficult to find somewhere to live at first.
2. doesn't need to speak the language of the country when studying there.
3. lives together with a lot of other foreign students.
4. studies in a country where they have small learning groups at university.
5. had to face a lot of official difficulties.
6. lives together with some other students.

R **c** **Answer the following questions.**

1. Why did Sven Berglund move out of the students' hostel in Warsaw?
2. What did Sven Berglund find strange at first and how did he solve the problem?
3. Why isn't it a problem for Brian Auger that his engineering lectures are in French?
4. What disadvantages are there if you study in France?
5. What benefits do you get in France if you insure with a students' insurance company?
6. Why isn't Esther Rosenkranz studying at Oxford University?
7. What differences are there between studying in Germany and Britain?
8. What does Esther Rosenkranz like about the hostel where she is living?

R **d** **Find verbs or adjectives in the text which go with the nouns in the box below. Say where you have found them and whether they are verbs or adjectives?**

Example:

I have found the word 'train' on line 2. It is a verb.

training • organisation • practice • insurance • application • payment • knowledge

importance • beauty • stranger • possibility • difficulty • excitement

2 The European Union

European unity has been a slow and sometimes difficult process. After the 2nd World War most people in Europe dreamed of a United States of Europe: democratic, peaceful and wealthy. However, many governments put their own political and national interests first. These factors have often slowed down the building of European unity. On the right you can read about some of the important steps on the road to a united Europe.

1957 Belgium, France, Germany, Italy, Luxembourg and the Netherlands …	… and Romania join the EU.
1958 The European Community …	… join the European Community.
1973 Denmark, Ireland and Great Britain …	… sign the Treaty of Rome.
1979 First direct elections for the European Parliament …	… become members of the EU*.
1981 Greece …	…welcomes 10 new members.
1986 Portugal and Spain …	… comes into operation.
1993 The twelve member states …	… joins the EU.
1995 Sweden, Finland and Austria …	… becomes the single currency of many EU states.
2002 The Euro …	… take place.
2004 The EU …	… introduce the Single European Market (The European Union).
2007 Bulgaria …	… enter the EU.

*"EC" before November 1993.

R **a** **Complete the sentences in the left column with the information in the right column. Use the Internet or any other source for help.**

P **b** Now give a report on the history of the EU in the past tense. Use the linking words on page 19 to join the sentences.

P **c** What has happened in the EU since 2004?

1 Travelling in Europe

Last summer Robert Page, a computer salesman, travelled around France by train. Here are his travel notes.

JULY

12 Monday
18.00 Paris Gare du Nord
evening - hotel (Montmartre)

13 Tuesday
morning - Eiffel Tower, Notre Dame
13.15 John Walters / business lunch
(Café des Artistes)

14 Wednesday
morning - shopping
13.10 train to Marseilles (Dan and Mike / workmates)

15 Thursday
morning - old port (very hot)
14.00 meeting / Martin Turner

16 Friday
11.00 breakfast / Hôtel du Roi, rest of day on beach
18.30 flight to Manchester

17 Saturday

a Answer these questions.

1. What time did he arrive in Paris?
2. When did he go to the hotel?
3. What sights did he see in Paris?
4. Where did he meet John Walters for lunch?
5. What did he do on Wednesday morning?
6. Who did he meet on the train to Marseilles?
7. When did he visit the old port?
8. What did he do on Friday morning?

b Make complete sentences with the following words:

Example:
after/arrive/Paris/go to/hotel
After he had arrived in Paris he went to the hotel.

1. after/be to/Eiffel Tower/meet/John Walters
2. after/meet John Walters/have/lunch
3. catch/train to Marseilles/after/do/some shopping
4. after/get on/train to Marseilles/meet two workmates
5. after/arrive/in Marseilles/visit/the old port
6. have/a meeting with Martin Turner/after/see/the old port
7. after/have/breakfast/spend/day on beach
8. catch/plane to Manchester/after/spend/day on beach

2 Some business trips do not go as well as others

Andy Williams met Robert Page and told him about what had gone wrong on his trip.

Use past perfect in one clause and simple past in the other. Use the words in brackets to join your sentences.

1. get to platform	train - leave already	(when)
2. the next train - be late	the engine - break down	(because)
3. at the hotel - find	leave my case on bus	(that)
4. get to bus station	it - shut already	(by the time)
5. ask a friend to meet me at hotel	he - not arrive	(although)
6. later - remember	give my friend the wrong address	(that)
7. leave the hotel to go for a meal	lose credit card	(after)
8. return to hotel	someone else - take my room	(by the time)

Example:
arrive at station - ticket office close already (when)
When I arrived at the station the ticket office had already closed.

3 At the airport

Mr Jones, a businessman from Cardiff, went to the airport last week to get his plane to Brussels.

a Make sentences about these situations.

Example:
drive to airport / start to rain
As/While he was driving to the airport it started to rain.

1. show ticket at airport / passport fall out of pocket
2. look around bookshop / meet a friend
3. plastic bag break / he and friend go upstairs
4. friend get coffee / he sit down at table
5. they call his flight to Brussels / he and friend drink coffee

b Jason Norman is a journalist for the "Evening News" in Seattle. Two weeks ago he travelled to Frankfurt. At the airport he went to the bank to change some money.
Put the verbs in these sentences into simple past or past continuous. Be careful, some sentences have both verbs in the simple past. Join your sentences with one of these words: *while, as, when* or *and*.

Example:
A lot of passengers (wait) at the counter - Jason Norman (walk) into the bank
A lot of passengers were waiting at the counter when Jason Norman walked into the bank.

1. he (look) at the exchange rates - man behind him (ask) to borrow a pen
2. he (stand) in line - he (count) his money
3. he (get) to the counter - Jason (hand) the money to the assistant
4. he (talk) to the assistant - there (be) an announcement
5. Jason (listen) to the announcement - the assistant (leave) the counter
6. she (return) - she (give) him a form
7. she (count) his money - Jason (sign) the form
8. she (give) him the money - he (put) it into his pocket

4 On the boat

Put the verbs in brackets into the correct form of the past tense (past simple, past perfect, or past continuous).

While Tim Smith **1. (stand)** and **2. (look)** over the side of the boat which **3. (take)** him to Ostend, he **4. (try)** to imagine the people who **5. (travel)** below him in the Channel Tunnel. Although the tunnel **6. (open)** many years ago and he **7. (have)** enough time to get used to the idea, he still **8. (find)** it difficult to imagine the journey to the Continent by rail and not by boat. Before they **9. (complete)** the tunnel, Britain **10. (always be)** an island and many people still **11. (want)** it that way.

1 Europe's cities lead the way

A long time ago, before people had even thought of ideas like the EU, Europe consisted of a number of independent nations. Nowadays, with more and more countries in the EU, the question is, what developments will tomorrow bring?

For some people, the most exciting aspect is its cities. And why? Well, look at any number of European cities even today and you can see that they are enjoying the greatest growth, energy and activity.
London for example, is exciting and the most cosmopolitan city in the world – the bridge between Europe and America. Berlin, on the other hand is the link between Western and Eastern Europe, its glass-walled Reichstag a symbol of the new European democracy. And it's not only in Europe's western capitals that you can feel this energy. The former eastern capitals – Prague, Warsaw and Budapest now not only look better after the neglect of the communist era, they also show a desire for the ideals of a modern Europe: tolerance and new ideas. Each city has its own strengths. Talinn in Estonia is rediscovering its strength as one of the great Baltic trading centres, while places like Marseilles, Barcelona and Liverpool are changing their run-down port areas into new centres which will link people to the sea. The diversity of each city is also one of its strengths. Each city is developing its own ideas. In Paris, a mayor once turned one of the main roads into a huge beach with sand and sunchairs instead of cars. The London Eye, also known as the Millennium Wheel, which was put up as a joke for the 2000 millennium, has now become a permanent attraction in London.
Even more exciting are the links between cities. High-speed trains and low-cost airlines are connecting more of these cities with each other. Such links mean that millions of Europeans can afford both the time and the money to visit places they could not have visited before. These links are also bringing Europe's cities closer to each other than to their own countrysides. Londoners will go shopping in Barcelona, while Parisians take the Eurostar to visit the free museums or the clubs in London. City leaders will talk to each other without waiting for the OK from governments.
In future Europe will have to look to its cities for new ideas, new jobs and new leadership. The growing links between cities will help to speed up the disappearance of national boundaries and each city and its people will gradually realize that their interests lie with each other. **431 words**

P **a** The photo shows something you can find in one of the cities mentioned in the text. → SF 19

1. Which city do you think it is?
2. What is the sight in the centre of the photo called?
3. Describe what you can see in the photo.

R **b** Welche der folgenden Argumente können Sie im Text finden? Zitieren Sie die zutreffende Textstelle.

1. Europas Stärke liegt darin, dass es aus unabhängigen Staaten besteht.
2. Die ständig wachsende Zahl der EU Mitgliedsstaaten wird die Wichtigkeit der europäischen Städte stärken.
3. Europas Zukunft spiegelt sich in den Städten, da sie die Orte des Wachstums sind.
4. In London leben viele Menschen aus verschiedenen Ländern.

5. Berlin ist eine wichtige Brücke zwischen Europa und Amerika.
6. Die Vernachlässigung der Städte im Osten Europas während der kommunistischen Herrschaft hat die Folge, dass sie jetzt eine nicht so große Rolle spielen wie die Städte im Westen Europas.
7. Als Folge der verbesserten Bahn- und Flugverbindungen sind jetzt immer mehr Menschen in der Lage, ausländische Städte zu besuchen.
8. Als Folge der Kontakte zwischen deren Einwohnern werden die Unterschiede zwischen den einzelnen Städten langsam verschwinden.

R **c** Answer the following questions on the text in complete sentences.
1. How has the view of Europe changed over the years?
2. What do London and Berlin have in common?
3. Why are Europe's eastern capitals also exciting?
4. What examples can you find of some new ideas?
5. What strengths do cities have individually and in general?
6. What advantages will new transport links between cities bring?

P **d** Project: Describe a European city and say what you like and dislike about it. Prepare a short presentation for your class (e.g. with the help of a presentation program). → SF 13+22

2 How Europeans go on holiday

I **a** Say which of the two photos below shows what kind of holiday you would prefer to take and say why.

Country	Holidaymaking rate (at least a trip of four nights)	Domestic trips	Outbound	Holidaymakers within the EU	Holidaymakers outside the EU
Belgium	40.2%	17.9 %	82.1%	73.9%	26.1%
Germany	76.9%	34.0%	66.0%	66.2%	33.8%
Spain	37.3%	89.7%	10.3%	59.4%	40.6%
Italy	46.3%	77.2%	22.8%	55.3%	44.7%
Netherlands	67.9%	37.0%	63.0%	74.7%	25.3%
Finland	51.9%	71.5%	28.6%	61.0%	39.0%
United Kingdom	60.9%	55.0%	45.0%	65.8%	34.2%

eurostat 2002

b Find out about the following: R
1. Which holidaymakers travel most/least?
2. Which holidaymakers stay in their home countries most/least?
3. Which nationalities travel within the EU/outside the EU most/least?

c Answer these questions. What do you think? R
1. Why do some people travel more than others?
2. Why do some people stay in their home countries, why do others prefer to travel abroad?
3. Why do many people stay within the EU, why do others take their holidays outside the EU?
4. Is holidaymaking different in central and southern Europe? Why/why not?

d Describe the main facts of the above chart in about 80 to 100 words. Use the results from exercises b and c. P

1 A gap year

a Listen to Andrea Miller from Leeds and find out if the following statements are true or false. Correct the false statements. R

1. Andrea returned from a gap year four months ago.
2. She travelled around for 10 months and also worked for a few weeks.
3. In the beginning she was having a hard time as her Swedish wasn't good enough.
4. Her job was quite boring.
5. She advised customers without any help.
6. After her internship she travelled to the southern part of Sweden and then to Oslo.
7. She wanted to travel longer but she didn't have enough money.
8. She is planning to get a job in a tourist information office.

b Would you like to study abroad? Find arguments for and against an exchange year.

2 Project work: Europe at a glance

The European Union consists of democratic European countries, committed to working together for peace and prosperity. Its intention is not to replace existing States, and it is more than just an organisation for international cooperation. The EU is, in fact, unique. Its member states have established common institutions so that decisions on specific matters of joint interest can be made at European level.

P Work with a partner or in small groups. → SF 13+22
Prepare and give a short presentation on one of the following EU institutions.
Use information from the Internet or from the media.

1. The European Parliament
2. The Council of the European Union
3. The European Commission
4. The Court of Justice
5. The European Economic and Social Committee
6. The Committee of the Regions
7. The European Central Bank
8. The European Investment Bank

Unit 4

Telecommunications/Advertising

Answer the following questions.

1. What can you see in the pictures?
2. Do you own or use any of the devices shown?
3. Are they useful?
4. What advantages do they have?
5. What developments do you think we will have in the future? Think of office technology and communications.

1 Your thumb hurts? – I know what you've been doing!

(A) My name is Sarah and I'm a text message junkie. I've known about my addiction for some time, but it's only now I realize that it's dominating my life. Call it what you will: text messaging, the text generation or thumb culture. Thankfully, I'm not alone. Millions of others – especially teenagers – in the UK send and receive text messages every day – even at school.

(B) Of course, it is big business for mobile phone operators and more and more marketing managers are now trying to enter the market for SMS (short message service) in order to promote their products or services. Broadcasters, newspapers and radio stations have launched competitions for SMS around shows or events. The results are amazing. The Sun, for example, recently ran an SMS competition called Ringo Bingo, which had more than 80,000 answers in the form of text messages.

(C) So where does SMS go from here? The cost of sending messages is going to come down. More and more promotions will use SMS; and it will likely move away from just being a young person's hobby to something that the older generations use. And, of course, thousands of people will make new friends via text message and thousands of others will break up in the same way. We seldom say what we really think face to face. It is easier to text a message when we want to express something private. And so even flirting becomes easier. What did we ever do without it?

(D) Education experts have warned that children's obsession with text messaging is seriously affecting their use of the English language. Increasing numbers of school children are using mobile-phone shorthand in written language. Traditional grammar, spelling and sentence construction are suffering as a result. But there are also a few teachers who claim that getting pupils to carry out written work using text language is better than nothing. I can understand the frustration of teachers and I think it's important that they get across to their pupils that text messaging is for fun but that learning to write proper English is more important for their future careers. Mark, 19 , from Swindon, a student in business studies says that he has been using text messages since he was 15. He adds that he has handed in essays that have been marked down for using text message shorthand. However, he also points out that it is really difficult to switch over to formal language when you have been texting for years.

(E) I'm managing my addiction well – I've only sent and received five messages while writing this piece.

415 words

(The Sunday Herald, 2003, adapted)

P **a** Describe the photo above. → SF 19

R **b** Refer to the photo and explain what the headline of the text means.

R **c** Match the following headings with suitable paragraphs from the text above.

> The future of SMS - Old people and SMS - The popularity of text messaging - The beginning of SMS - The author's addiction - The SMS business - Text messaging and the language

R **d** Find words in the box that go together according to the text.

> children - competition - English - message - mobile - off - phone - proper - school - SMS - switch - text

e Geben Sie kurz sinngemäß auf Deutsch wieder, wie sich der Einsatz von SMS bei jungen Leuten möglicherweise auf die Qualität der englischen Sprache in der Schule auswirkt. M

f Do a survey in your class and prepare a report about it. How many students P

1. … own a mobile phone?
2. … use it every day?
3. … do text messaging? (How many messages a day?)
4. … use text message shorthand?
5. … have taken part in competitions via SMS?

g Comment: Will text message shorthand have influence on the way we communicate in the future? P

2 Who is in touch?

The bar graph shows the percentage of electronic communication devices in five European countries.

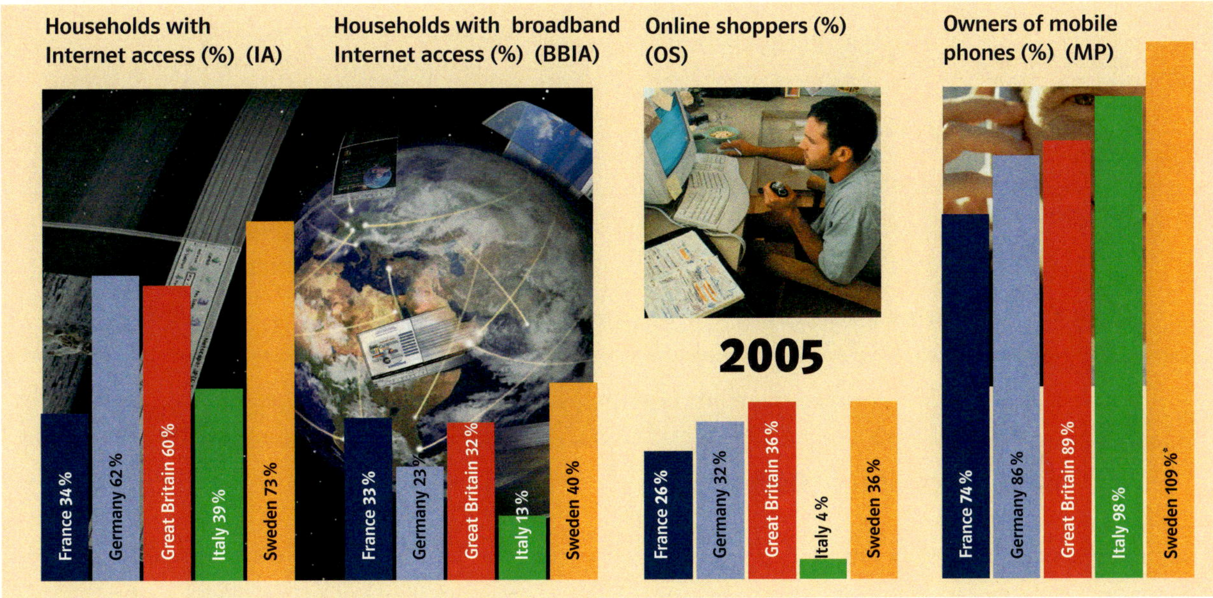

eurostat, May 2006

P **a** Compare countries in the following way. Use the phrases in the box on the right.

Example:
IA – S / GB
The graph shows that the percentage of homes with internet access in Sweden is higher than it is in Britain.

1. BBIA – I / S
2. MP – GB / G
3. IA – F / I
4. OS – S / I

The graph shows The figures show We can see from the statistics	that	the number of the percentage of	people households homes who…	is higher is lower	in (country A) than it is in (country B).

P **b** Describing a chart: ➜ SF 21
What does the chart tell you about the use of communication devices in Germany and Italy?

P **c** What could be the reasons for these differences? Use the following phrases.

I'm pretty sure that …
That may be because …
A possible reason is that …
I'm sorry, but I've no idea why …

1 Life in the future – what will it be like?

a Ask your classmates what they think about the following ideas about the future. The words in the box below can help you.

Example:
unemployment / increase
What do you think, will unemployment increase?
I think unemployment will increase.

I'm of the opinion; I think/don't think; I expect; I suppose

1. people / read / books any more
2. computers / control lighting, heating and cooking
3. we / have / more wars
4. television, DVD, and computer games / replace / sport
5. there / be / more pollution
6. all houses / use / alternative energy

b Now answer the questions in short form and give reasons.

Example:
What do you think, will unemployment increase? Yes, it will because computers will do all the work.

2 In the office of the "Daily News"

Make sentences like this:

Example:
The phone is ringing. A secretary is walking towards it. She is **going to** answer the phone.

1. Jim Price has sat down at his computer. His notes for a story are in front of him.
2. Sally Jones is looking at her address book. She picks up the phone.
3. Two secretaries have left their desks and are putting money in the coffee machine.
4. Sam Smith, the head reporter, has invited a local politician to the office. He is preparing some questions.
5. It is eleven o'clock in the morning. Jim is taking some biscuits out of his bag.
6. The receptionist and her friend are in the restaurant. They have chosen their food and are standing at the till.

3 Office Electronics Ltd

Sam Smith wants to write an article about developments in the modern media for the "Daily News". He has invited James Bradshaw, manager of "Office Electronics Ltd", for an interview. Here is part of their conversation.

Put the verbs into the correct form of the future.

Sam Smith: Come in, Mr Bradshaw, sit down. **I'll take** your coat and I expect **Janet will** bring us some coffee in a minute.
James Bradshaw: Thank you, Mr Smith.
Sam Smith: Now, as I said on the phone **I'm going** to do an article on media developments and I'd like to ask you a few questions. Don't worry about that noise, by the way, it's one of the old fax machines. **It's going** to break down any minute.

James Bradshaw: Oh dear, I hope it's not one of ours.
Sam Smith: Oh no, of course not. Now, Mr Bradshaw, how is business? Do you think this year **1. (be)** as successful for you as last year?
James Bradshaw: Well, I certainly hope so. I have our most recent sales report here and the figures show that the firm **2. (do)** very well again this year.
Sam Smith: Of course, the market is growing all the time, isn't it? In the next few years I expect we **3. (see)** an even bigger increase in the use of communications technology. But do you think this **4. (be)** a good thing?
James Bradshaw: Well, for me and my company certainly!
Sam Smith: But what about the customers? It's an expensive business, isn't it? I've certainly decided that I **5. (not spend)** any more money on the latest developments.
James Bradshaw: Well, just a minute. I **6. (show)** you our brochure …

4 The trainee journalists

Two new trainee journalists have started work at the "Daily News" newspaper in London. They are on a one month test period. Ken Brown, the head of the foreign news department, has to write a report on them. He has used the following system:
++ excellent / + good / – bad / –– terrible

	General work	Computer skills	Shorthand	French	Communication skills
Duncan	++	+	–	++	––
Tina	+	–	+	–	++

a First make sentences about Duncan. Use the verbs in the box.

speak – use – communicate – write – work

Example:
Duncan's general work is excellent. He works excellently.

b Form adverbs from the following adjectives.

careful, slow, fast, accurate, polite, quick, correct, hard, wonderful

c Now choose from these adverbs to compare the trainee journalists in the following way:

Example:
general work
Duncan works more carefully than Tina does.
Tina doesn't work as carefully as Duncan does.

5 A report

Look at the adjectives and adverbs. Then complete Ken's report on two other trainees.

On the whole I can say Phillip's work in the office has been **fairly good**. **Unfortunately**, he speaks French **terribly**. On the other hand, his computer skills appear to be **very good**. In comparison, Charlotte has worked **extremely well** over the last four weeks. Although she speaks French **(real) (bad)**, her computer skills are **(extreme) (good)**. Nevertheless, she seems to be the **(good)** of all our trainees. **(General)**, she keeps **(calm)** in a crisis and doesn't get **(nervous)**. All in all, I think she will **(probable)** make an **(excellent)** journalist. But she must learn to speak more **(polite)** on the phone. I **(certain)** cannot say the same about …

Development of advertising

1 Change in share of advertising 2000–2004

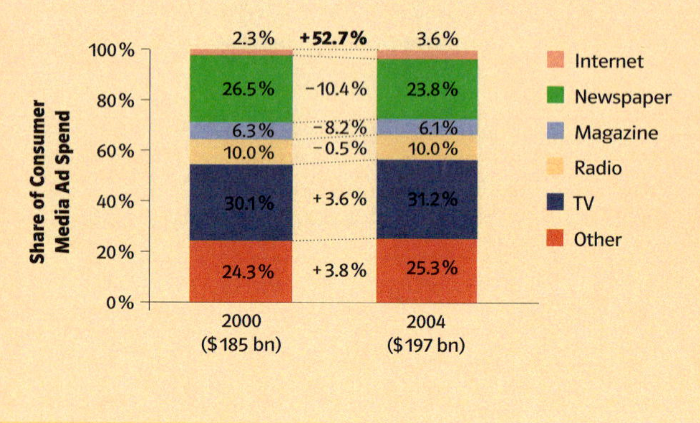

(Source: www.doubleclick.com)

2 The growth of online advertising

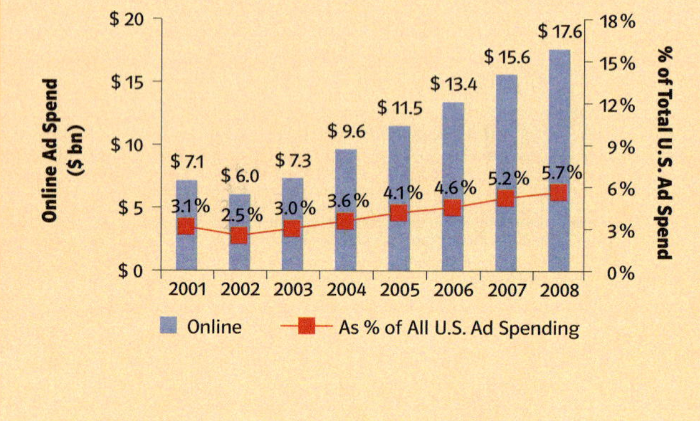

(Source: www.doubleclick.com)

P **a** Compare the advertising market regarding the share of the different media in 2000 and 2004. (Chart 1) → SF 21

P **b** Describe the development of online advertising between 2001 and 2008. (Chart 2) → SF 21

R **c** Read the text on the next page and answer the following questions.

1. Which of the media in chart 1 above are mentioned?
2. What could the headline of the following text be?

If you see a big yellow M on TV, will you recognize it and – more importantly – will you know what product it stands for? Most people will recognize it, according to an international survey. They believed that the M was one of the best known logos in TV advertising.

For the companies that had spent millions on the production of commercials and on airtime the results of the survey were quite important. The companies whose products came first were pleased. But the problem is that compared to every commercial that becomes popular 100 others fail.

Indeed, according to the survey, one of the main disadvantages of TV advertising is that only a third of commercials receive our full attention. One in five plays to empty rooms, while people are making tea or going to the toilet. Two in five act as background while people are eating, reading or even sleeping. With the help of "hidden" video cameras in a number of homes, the survey showed what surprising things people actually do during the break – one man even practised his tennis serve.

According to the survey even people who stay in the room during commercials do not always watch them. A new generation of viewers has arrived – people who watch with remote control in hand and switch to another channel when the commercials come on. Reports from America, where this habit is much older, say that in this way companies waste a quarter of the money which they spend on TV advertising.

Of course, it was not the TV companies which paid for this international survey, it was the newspapers. They used it for an £8 million advertising campaign. The message was simple: Advertisers overestimate the value of TV commercials. The reason why newspapers started their campaign is also easy to see; their share in the advertising market has been decreasing worldwide, and is down to about 15 % in Britain.

Newspapers and TV stations have different viewpoints on advertising. On the one hand, Bert Hardy, who is the manager of the newspapers' campaign, argues: "The most expensive time in TV advertising costs around £60,000 a minute, whereas a full page in a national newspaper only costs £20,000 and lasts all day. The companies could use some of the money that they waste on TV advertising to increase their profits and cut prices." On the other hand ITV manager Malcolm Wall points out the fact that newspapers themselves spend millions on TV advertising.

Today many advertising managers are sure that a mix of the advertising media will be the best way to reach consumers. Some companies have already changed their advertising campaigns. Many more are planning the same and are going to advertise not only on TV in the future but also on the radio, in newspapers and on the internet.

465 words

(The Telegraph plc, London)

R **d** Find the expressions with the same meaning in the text.

1. 20 per cent 2. 25 per cent
3. 33 per cent 4. 40 per cent

P **e** Try to make some more expressions in the same way.

1. 66 per cent 2. 75 per cent
3. 10 per cent 4. 80 per cent

R **f** Short answer questions/Sentence completion: Answer the following questions by providing the required information or by completing the sentences, both with words from the text.

1. Which verb in the text shows that not all companies are happy with the results of the survey?
   Many advertising campaigns …
2. What is the share of viewers that really watch commercials?
3. Who paid for the international survey on TV commercials?
4. Which appliance helps you to choose different channels on TV?
5. Which advantages of advertising in newspapers are given?
   Advertising in newspapers is less … and it … the whole day.

g Fassen Sie kurz den letzten Abschnitt auf Deutsch zusammen und erläutern Sie, welche Werbemedien in Zukunft eingesetzt werden. M

h Give your opinion. P

If advertising on TV is so expensive, why are firms willing to spend so much money on it?

1 Radio advertising

a Listen and find out: R
1. the product
2. the name of the product
3. the price
4. the name of the shop that sells it

b Listen again and find adjectives from the ones below which describe the products. Say which of the products they describe. R

careful • cheap • superfast • superb • modern • safe • excellent • popular • polite • wonderful • light • accurate • good • quick • easy to use • comfortable

2 Analyzing advertisements

a Try to analyze the advertisement on the left. P
1. Say what this advert is for.
2. Identify
   - the slogan/key words.
   - the logo/brand name/name of the company.
3. Say what other information is given in this advert.
4. Describe what the picture shows.
5. Say what ideas and lifestyle you associate with this picture.
6. What target group could the advert be aimed at?

b With reference to what you have found out above do you think this advert could be successful?

3 Writing an advertisement

a Look at the photo on the left and think of a product/service you could sell with it. Choose a suitable target group.

b Now write a short advertisement for your product/service. Refer to the points in D2. P

c Present your advertisement to the class. → SF 13 P
Say why you have designed it in this way.

4 Designing an advertisement on the PC → SF 13+22

a Work in teams. Decide on a product or service you would like to sell.

P **b** Draft an advertisement for a men's / women's / children's magazine. Use suitable digital photos or collect photos from the Internet. Make up a slogan, a logo and necessary information on your product or service. Prepare your presentation with the help of a presentation program.

P **c** Present your advertisement to the class.

d After the different teams have presented their ads, decide which one your team likes best. Give reasons.

Transport

P **a** Describe the cartoon above and point out its message. → SF 20

b Decide which of the following comments in your opinion best describes the situation in the cartoon and say why.

1. People in cities don't need their cars.
2. Private transport will be the death of us.
3. If you want to reduce your stress level, leave your car at home.
4. It's quicker on foot.

c What is your view of the traffic situation in our towns and cities?

1 Traffic

Every day we can hear reports of cities which are congested by traffic and this situation is getting worse day by day. The congestion of roads and motorways causes major problems for drivers and the environment. These problems have grown with the development of private transport. We have got used to a lifestyle we find hard to change. More and more people who own cars are not able to imagine a life without them. The question is how to find a solution people would accept. Some people believe public transport should be the answer. Others are looking into technological innovations such as electronic navigation systems. **109 words**

a Match the following words with their definitions. **R**

1. congestion 2. environment 3. navigation system 4. public transport	A. the world around us B. a way to travel for people in large groups, e.g. bus, tram, train C. the state of being too full or overcrowded, especially on roads D. a device which helps you find your way

Here are some statements we heard at a transport planning meeting.

b Complete these statements with the words used above (1.–4.). **R**

1 *"During the rush-hour the … on the M25 leads to long traffic jams every day."*

2 *"The fumes from cars pollute the …"*

3 *"A … can help drivers to avoid congested traffic without having to look at a map."*

4 *"I believe we should use … and leave cars at home."*

2 Car navigation systems

Car navigation systems are no longer a high-tech dream. You can find them in many cars as a standard device. If you miss a traffic report, there is no need to worry. Your navigation system saves it for you in text and audio form and will give it to you at the touch of a button. It can even recalculate your chosen route using the latest traffic information. Life couldn't be easier!

Drivers can relax as they are guided to their chosen destination, be it a city centre or a small suburban street, without having to look at a single map. A quick look at the display and you can see how much longer your journey will take and how many miles you have to drive. All this is made possible by satellites that can pinpoint your car's location if it is equipped with a Global Positioning System (GPS).

The Trafficmaster system is an additional aid in avoiding traffic congestion. Instead of relying on radio reports, Trafficmaster obtains its information from sensors installed on motorways and road networks which monitor traffic speed. Trafficmaster transmits this information to its users at regular intervals. The system is available as part of the navigation system from some car manufacturers, but is not normally a standard feature and you have to buy it as an extra.

So, is there anything navigation systems cannot do? Well, they cannot regulate traffic flow to avoid traffic jams and slow moving traffic in the first place. They can only help find a way out of the congestion and suggest a diversion. The other disadvantage with most navigation systems at present is the slow and rather complicated method of entering your destination, especially when you are driving. However, this is all going to change. A leading technology company and a car manufacturer have produced a navigation system that uses voice recognition. For example drivers who suddenly realize that they are running out of petrol can press the 'talk' button on their steering wheel and give commands like "find the nearest petrol station". The system is programmed to understand around 150 commands and a number of different accents. This makes programming your system a lot easier and safer, for example, when you are driving along the motorway or looking for the nearest fast food restaurant. However, all this only helps the individual driver and does not really solve the general problem of traffic congestion and pollution.

406 words

R **a** Look at the text.

1. Describe how you can find where the nearest petrol station is through a car navigation system.
2. Explain how this is possible.
3. Show how the Trafficmaster system works.
4. Point out two disadvantages of car navigation systems.
5. Explain why the newer navigation systems are safer.

R **b** Here are some radio travel news announcements. Find nouns in the text to complete them.

1. Traffic … on the M1 near Nottingham is causing problems for drivers. Try to avoid the M1 near Nottingham if possible.
2. Due to motorway repairs on the M4 there is a 5-mile … in both directions near Bristol.
3. Road works on the A58 south of Rochdale. The police advise drivers to follow the … signs.
4. Due to an accident on the A30 near Exeter there is 3 miles of … So take care when driving on the A30 near Exeter.
5. Fog on the M62 ! Drivers should reduce their …

P **c** Rewrite lines 10 – 18 and replace some of the expressions with those in the cloud. Start like this:

Example:
Drivers can take it easy as …

monitor • **take it easy** • route • glance • just one • find exactly • how far • fitted • position • shown the way

1 Getting around

a Look at these pictures and the information below. Then make sentences like this:

Example:

1. The train which is standing at the station is going to the Docklands.
2. The students who are waiting for the ferry …

- go to the Docklands
- go on a bike tour
- phone his wife
- be a London tourist bus

b Now describe the other pictures in the same way. → SF 19

Example: The train which is going through the Eurotunnel …

- go to Paris
- travel to Italy
- read newspaper
- go to Paddington

2 Travelling home

Complete the following story. Fill in "who" or "which" where necessary.

At 5.30 Henry arrived at the station **(–)** he knew so well. He showed his ticket to the man **who** was standing at the gate. Then he got on the train **which** was waiting at platform 7. He politely greeted the woman **(1)** was sitting next to the door. She put down the book **(2)** she was reading. They both smiled and then he walked to the seat **(3)** he always chose. He sat down and filled the pipe **(4)** he had taken out of his jacket. This was the time of day **(5)** he liked best. After some minutes he started to read the paper **(6)** he had bought at the station. The couple **(7)** were sitting next to him were talking about the plans **(8)** they had for the weekend. He listened for a while and then he looked out of the window at the countryside **(9)** was passing by. What a nice day it was! Henry started to think about the work **(10)** he had done at the bank. He smiled. He remembered all the years **(11)** had passed and all the people **(12)** had worked with him. When the train arrived at the station he at once saw the woman **(13)** was waiting for him. He walked up to her and kissed her. It was the first time his wife had met him at the station. She had bought him the flowers **(14)** he liked so much and he knew it would never be the same again. This had been his last day at work after 48 years.

3 A guessing game

a Guess who or what it is.

1. It's a person who goes on a journey for pleasure.
2. It's an electronic appliance which helps you to find your destination.
3. It's a place where trains arrive and leave.
4. It's a piece of paper you buy when you want to travel by train or bus.

b Now describe other persons, places, vehicles etc. in the same way. Don't mention their names. Your classmates should find out who/what they are.

4 Changes in travelling

■ Compare travelling today, in the past, and in the future. Make sentences like this:

Example:

Lots of people must/have to wait in traffic jams today. 100 years ago most people had to walk.
In 100 years' time I think people will …

Today …	100 years ago …	In 100 years' time (I think) …
lots of people/wait/ in traffic jams (must/have to)	most people/walk	people/use/public transport in cities
passengers/travel/to America in a few hours (can/be able to)	passengers/get/to America in a week	passengers/travel/around the world in minutes
drivers/stop/at most European borders (needn't/not have to)	drivers/have/a driving licence	drivers/stop/at any borders
travellers/enter/some countries without showing their passports (can/be allowed to)	travellers/take/their cats and dogs into Britain	travellers/enter/all countries freely
man/fly/to Venus (can't/be not able to)	man/fly/to the moon	man/fly/to the stars
people/drive/ faster than 70 mph on British motorways (mustn't/be not allowed to)	people/drive/without a person carrying a red flag in front of their vehicles	people/drive/cars in city centres

5 A day in the life of Harry S.

Harry **(1)** works in London. But he lives in Bacon End **(2)**. Every day he has to travel one and a half hours to the computer firm **(3)**. At 7.30 when he leaves his house, he gets into the car and drives to the station **(4)**. Harry arrives there at about 8 o'clock and he gets on the train **(5)**. On the train he reads the newspaper **(6)**. Sometimes he talks to the passengers **(7)**. When he arrives in London he takes the underground line **(8)**. Sometimes he meets his colleague **(9)** and together they walk to work. On his journey home **(10)** he often uses his mobile phone to text his friends or listens to his CD player.

■ Complete the text about Harry S. with the information below. Use relative clauses. Decide if the relative clause is necessary or not and find out if you can replace "who" or "which" by "that".

Example:

Harry, who is 24, works in London. But he lives in Bacon End, which is 35 miles away.
Every day he has to travel one and a half hours to the computer firm he writes programs for. At 7.30 …

Some more information on Harry S.:

a) The journey home takes about one and a half hours.
b) The line goes to Cannon Street Station.
c) Harry is 24.
d) He writes programs for the computer firm.
e) He usually buys a newspaper at the station.
f) Bacon End is 35 miles from London.
g) The passengers are sitting next to him.
h) His colleague also works at the computer firm.
i) The station is in Harlow.
j) The train leaves at 8.05.

1 From jam to tram

(A) Peter Wood used to travel to his office in Manchester by car. Every morning the journey took longer and longer because the roads were too congested. One day he decided to try the tram and the journey was half an hour shorter. Thanks to the town's tram system, his journey only takes 16 minutes. "It really is a very good service," says Wood. Other commuters are happy, too. One of them, Mike Pulman, adds "I stopped using my car years ago simply because the tram is so convenient."

(B) Their views of the tram are quite different to the ones people in Britain had after the war. At that time people thought that the trams were noisy and slow and so they got rid of them. In the 1920s, Britain had over 14,000 trams in more than 100 towns, but by the 1960s the tram remained in only one town. At that time the attractive alternatives were the bus and the car. The bus because it provided a more flexible alternative to the fixed-route tram and the car because of the freedom it gave to the individual driver. Of course, in those car-crazy days, "nobody thought about the congestion and pollution which all those petrol engines would produce," says Don Roach, who is head of London's traffic department. Today however, environmentalists and city planners are welcoming back the electrically-driven, new-style tram or 'light-rail system' as they now call it. They see it as an efficient and ecological solution to inner city traffic chaos.

(C) Today's trams are most profitable in cities which have populations of half a million or less and no more than 20,000 passengers an hour at peak times. Although one alternative – an underground system – may be a good solution to the traffic problems of large cities, Mary Browning, a British environmentalist says "it doesn't make sense for medium-sized cities and is much more expensive to build than a tram system." In fact, a team of traffic experts has estimated that building one kilometer of underground costs 11 times more than building the same length of tramway. "The main reason trams are coming back is not just because they are good for the environment," says Browning, "it's because they are cheaper and more efficient."

(D) However, the fact that there are two good arguments for trams doesn't mean that they are easy to introduce. Public transport of any sort is unpopular with car drivers and the car industry so that governments do not always support it. As a result, it is difficult for town planners to put tramway projects into practice. This helped to slow down the opening of Manchester's system by nine years and may also slow down progress on similar projects. However, since statistics show that there will be twice as much road traffic by the year 2025, sooner or later governments will have to look at light rail systems as one possible solution to the transport problems which European cities are facing in the 21st century.

498 words

R **a** Find the forms of transport mentioned in the text.

R **b** Match four of these headings with the paragraphs in the text.

Trams – an old idea made new • Trams – a danger to private transport? •
Trams reduce journey time • Trams – too noisy and dirty •
Trams save money for medium-sized towns

P **c** Now do these tasks.

1. Describe the difference which Manchester's tram system has made to commuters.
2. Say why trams were unpopular in Britain after the war.
3. Explain the advantages which today's modern 'light-rail systems' have.
4. Compare the costs of a tram system with an underground system.

P **d** Complete these sentences with information from the text. Use your own words.

1. As a result of road congestion in Manchester commuters …
2. Traffic planners are in favour of the tram because …
3. Although Britain had 14,000 trams in the 1920s …
4. Buses became more popular than trams because …
5. Although there are two good arguments for trams …
6. As a result of the increase in road traffic in the future …

M **e** Erläutern Sie, warum es schwierig ist, in GB Straßenbahnsysteme einzuführen. → SF 10

2 Solutions to traffic problems

Two traffic experts are talking about possible solutions to traffic problems:

A: But **don't you think the best thing would be to** improve the public transport system?
B: Certainly, but that's not enough. People will never give up private transport completely. They would lose their freedom and industry would suffer. **I think the answer to the problems of** congestion for example, is technology. Why not make car computers cheaper, then everyone could have one?
A: Yes, but that doesn't solve the problem of pollution, does it? **One solution to that problem could be** a tram system. It's the quickest and most convenient means of transport and **I would definitely suggest it** for a medium-sized town.

I ■ Discuss the solutions you can see in the pictures below in the same way and any others you can think of. Use the expressions in bold type in the dialogue or similar ones. → SF 14

1 Travel announcements

a Listen to the announcements. R
Where could you hear these announcements?

b Listen to the announcements again and and find out if the following statements are true or false. Correct the false statements. R

1. The train to Liverpool leaves at 10.30.
2. Passengers for Manchester have to change at Birmingham.
3. The train does not stop in Leeds.
4. Flight BA 98 to Dublin is ready to take off.
5. Passengers flying to Newcastle have to go to Gate A 6.
6. Bad weather has led to a delay.
7. There is a hold-up on the M4 motorway near Birmingham.
8. The eastbound section of the motorway is closed because of an accident.
9. Drivers should use a detour route.

2 Welcome to Wakefield

It's so simple to get here!

Since the great coaching days of the 18th century Wakefield has always been at the "crossroads of the north". This fact is now even more important in a world where fast and easy communications are a must. In any part of the Wakefield district you are only minutes away from a motorway, the M62 to the two main ports of Hull and Liverpool, and the M1 to London which is only a three-hour drive away.
Moreover, you can reach Wakefield by rail from all parts of the country. Regular train services operate from most cities in England, Scotland and Wales. Again London is only two hours away.
On top of all this, Wakefield has two major airports nearby: Manchester (only an hour's drive on the M62) and Leeds (only minutes away via the M1). These fast growing airports provide links with four major intercontinental airports – London, Amsterdam, Paris and Dublin.
An ideal base. Discover Wakefield yourself. It could work for you.

For further information write to:

Wakefield
Tourist Information Centre
Town Hall
Wood Street
Wakefield WF1 2HQ

Or visit our website:
www.city-visitor.com/wakefield

162 words

R **a** With the help of the text and the map answer these questions about Wakefield.

1. Why do you think Wakefield is called the "crossroads of the north"?
2. How can you get to continental Europe from Wakefield?
3. Why are the Manchester and Leeds airports so important for Wakefield?
4. Why do you think Wakefield has produced this promotion brochure?

P **b** Design a promotion leaflet in English for your town. → SF 13
Then present your leaflet to your class.

Unit 6

Environment

Waste management

In one year the average family of four throws away 112 pounds of metal, 90 pounds of plastics and an amount of paper equal to six trees. Each person in the EU produced around 580 kilogrammes of municipal waste in 2003. Where does all this waste go?

Municipal waste management in the European Union 2003

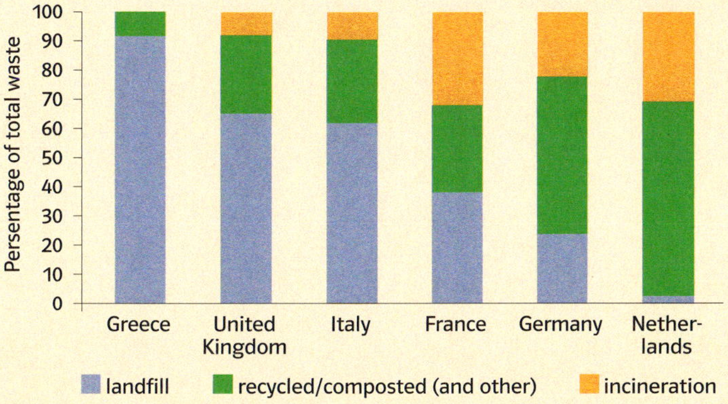

a Talk about the picture above and what you can see in it. → SF 19

b Answer the following questions. R

1. What kind of waste do you think it is?
2. Where, in your opinion, does this waste come from?
3. What will happen to the waste?
4. What else could they do with it?

c Describe the information in the chart on the left. → SF 21 P

d What are the current developments in waste management in Europe, in Germany and in the area where you live? Compare the latest figures (e.g. from the Internet) with the figures in the chart. P

1 Everyone can help to save the world

(A) We are all environmentalists now. We all know about acid rain and the ozone hole. But what are we doing about these problems? As consumers we have the power to make decisions which will contribute to a better world. For example, when we walk into a supermarket we can choose to buy environmentally-friendly products and refuse products which damage our planet. In this way we also have the power to influence retailers to sell us these so-called 'green' products.

(B) The supermarkets have understood the message and are now trying to improve their environmental image. They are turning to less wasteful and recyclable packaging. The range of 'green' products will continue to expand to meet the growing demand. This message has also reached the manufacturers. Companies are now developing environmentally-friendly products which mean no decrease in quality or increase in price. They have realized that protecting the environment may start as an advertising campaign but in the hands of intelligent people it can become a very profitable business.

(C) We also know there are other areas where all of us can make a positive contribution to a cleaner, healthier and safer environment. We can all reduce energy consumption at home and so cut the emissions from power stations. There are a number of ways we can do this. For example, we now know how important it is to insulate our houses well. In the use of electrical appliances and lighting we should choose the best energy-efficient technology, such as modern light bulbs which can reduce electricity consumption by 80% over the older ones. Rubbish is another area where we can act. We all produce tonnes of rubbish and so waste disposal is becoming a big problem. As an environmentally-conscious consumer we should buy products which do not have so much packaging. We should avoid non-returnable goods at all costs. In Scandinavia and Germany, for example, plastic bottles and cans are now returnable goods.

(D) In conclusion it is important to emphasize once again the rise of consumer power. We must realize that we are all part of the problem. Nevertheless we must also realize that we have the responsibility to be part of the solution.

363 words

R **a** The text above has four paragraphs. Find a heading for each paragraph in the following bubbles.

- How we can influence corporations
- The use of public transport
- Everybody's duty to contribute to a cleaner world
- The introduction of 'green' products and price increases
- The problems of acid rain and the ozone hole
- Other ways to protect the environment
- The reaction of retailers and manufacturers to environmental problems

b Answer the questions on the text. R

1. Which decisions can a consumer make in a supermarket?
2. How have supermarkets reacted to these decisions?
3. Why have manufacturers become more interested in environmentally-friendly products?
4. How can we reduce our consumption of energy?
5. How can we help to avoid waste?

c Give your opinion.

1. What is the message of the text?
2. In what other ways can you protect the environment?
3. What other examples of 'green' products can you think of?

d Look at the following nouns and find the verbs in the text. Say which line in the text.
1. choice
2. reduction
3. insulation
4. production
5. emphasis
6. sales

e Now find the nouns from the following verbs in the same way.
1. to consume
2. to demand
3. to decide
4. to contribute
5. to waste
6. to solve

f Find synonyms in the text for the following words.
1. shop owners
2. 'green' products
3. waste
4. to manufacture
5. to cut
6. to make better

2 Pollution

P **a** Describe the photos on the right. → SF 19

R **b** Read the statements and match them with the photos. One photo does not fit.

A Emma Shaw, chemist: "I've just read a book about acid rain. Did you know it was a Scottish chemist, Robert Angus Smith, who invented the expression 'acid rain' in an 1852 article on the air and rain in Manchester? Moreover, according to him air pollution damages buildings and forests more quickly in countries where people use more fuel than elsewhere."

B Paul Jenkins, environmentalist: "It's true that this government is doing a lot to reduce the level of air pollution here in Britain. However, in my opinion, we consumers have the power to make the biggest improvement. We should buy cars which don't pollute the environment so badly or we should use public transport."

C Jill Kemp, journalist: "As I see it, we, the people of Britain, have been using our rivers as waste disposal channels for hundreds of years. I'm convinced that our government must act. Furthermore, we need better pollution control over companies, farmers and even local government."

D Richard Palmer, politician: "I think the ozone hole is the greatest danger to our planet. It isn't true that we can afford to wait. The ozone hole will cause serious dangers to our health. In addition all over the world people must stop making household products with CFCs* in them."

* CFC = chlorofluorocarbon

1

2

3

4

5

c Why did you make those decisions? Answer in the following way:

Example:
I chose the photo of … for text A because Emma Shaw says/tells us …

P **d** Report what each person says and introduce each reported sentence with a different verb. Use the table below.

	Direct speech	Reporting verbs
Opinion	"In my opinion …" "As I see it …" "I think/believe …"	be of the opinion take the view think/believe
Emphasis	"I am (I'm) convinced …" "We should/ought to/must …"	maintain emphasize
Agreement/ Disagreement	"It is (It's) true …" "It is not (It isn't) true …"	agree not agree
Additional Information	"Moreover …" "Furthermore …" "In addition …"	point out go on to mention add

1 Protect your environment

For a project on the protection of the environment, the consumer magazine 'What?' interviewed people in Birmingham. Here are some of the answers they got.

1. "We are fighting for more cycle paths here." **Joe Straw, member of Green party.**
2. "I'm joining the WWF tomorrow. People don't give enough money to help with their work." **Roger Williams, bank clerk.**
3. "I don't buy cans." **Mrs Black, housewife.**
4. "I ride my bike to school every day." **John Smith, teacher.**
5. "We are closing some streets to traffic in the city centre next year." **Ray Peters, town planner.**
6. "My husband doesn't wash his car here in the street any longer." **Mrs Rivers, housewife.**
7. "We always buy bottles we can return to the shop." **Fred and Jim, college students.**
8. "Well, I'm just taking my old newspapers to the container like I usually do on Saturdays." **Peter Bone, old man.**
9. "We don't buy goods which aren't environmentally-friendly." **Anna and Rob Nelson, young married couple.**
10. "We must all save a lot more energy in the home. We're buying an energy-efficient heating system tomorrow." **Mr Field, farmer.**

■ Report the answers like this:

Example:

1. Joe Straw, a member of the Green party, said that they were fighting for more cycle paths there.
2. Roger Williams, a bank clerk, said that he was joining the WWF the next day. People didn't give …

2 Retailers and the environment

The manager of a large supermarket group gave the following statement to 'What?' magazine about retailers and the environment:

"We opened our first supermarket just ten years ago and we have had a lot of success since then. We have always kept our prices low and our quality high. We will, of course, continue to do this in the future. However, and I know you are interested in this, we have always had a special interest in environmental matters. Right from the start we sold goods with as little packaging as possible and placed bottle banks and containers in all our car parks. As from next month we won't hand out free plastic bags any more and we will stop selling all drinks in plastic containers. Over the last few years we have introduced more and more 'green' products and we certainly haven't felt sorry about this. Two years ago, when we started to plan our second store here in this area, there wasn't a bus service. But after we wrote to the council they agreed to introduce one. All in all I think I can say that we are really doing our best for the environment."

Chris Hawkings, who reported this statement for the magazine, began his article like this:

Example: The manager of Dippy Supermarkets spoke to me yesterday … He told me that they had opened their first supermarket ten years before and that they had had a lot of success since then. He went on to say … that they had always … and that they would of course …

■ Complete his article.

3 Interview with "Whiz Bang Electronics"

Chris Hawkings' next job was to interview the head of "Whiz Bang Electronics" about their new washing machine, which they want to sell as an environmentally-friendly model. Here is his list of questions:

1. Do you plan to replace all your old models with environmentally-friendly ones?
2. How many models will you produce per year?
3. Why do you call it environmentally-friendly?
4. Do you think the new model will sell well?
5. When did you begin to develop this model?
6. How many litres of water does it use?
7. What is the energy consumption like?
8. What plans have you made to recycle these washing machines?

Later at lunch, the head of "Whiz Bang" told one of his colleagues about the interview and that the reporter had wanted to know a lot about their new model.

Example:

1. He wanted to find out if we planned to replace all our old models with environmentally-friendly ones.
2. He also wanted to know how many ...

■ Now you go on.

4 The inner city traffic situation

Last night there was a meeting in the town hall to discuss the inner city traffic situation. Here is part of this discussion:

Town planner: ... and so in my opinion, the only solution is to close the centre to traffic completely.
Shop owner: What will happen to our trade? No one will want to come into town at all then. And we already have enough competition from these big out-of-town superstores.
Town planner: We have introduced a number of highly successful park and ride schemes and I'm sure people will prefer shopping without all the noise and exhaust fumes from cars.
Shopper: That may be true, but we don't like carrying heavy bags of shopping for miles and miles before we can get them into the car.
Shop owner: You see, we must discuss the problem in more detail. Do you realize the problems we will have with deliveries?
Town planner: We do, Mr Black, and as long as lorries can deliver before 10 in the morning, there won't be a problem.
Shopper: I suppose shopping is more pleasant with no traffic. I will certainly feel better about bringing my children into town then. In some cities they even have special play areas for children. Can't we have something like that here, too?

■ Now report this discussion. Use as many different reporting verbs as possible.

Vehicles go round again

The three R's of waste management

Randy Newbury, manager at an international car company, has a dream. "In thirty years our company will be 100% environmentally friendly. On the one hand we will not pollute air or water any more during the production process, on the other hand we will only produce cars from recycled material. And I hope we will then be able to recycle these cars completely." This sounds unlikely? Not to Newbury, who is responsible for such environmental problems. What is more, lots of other international companies are actively trying to find solutions to make this dream come true. While they are already re-using waste heat and raw materials, they are reducing the amount of waste in the production process. Moreover, they are trying to increase the number of parts they can re-use. And of course, the products themselves must be recyclable. To be more precise, these companies are employing the new 3 R's to protect the environment: reduce, re-use, recycle.

The 3 R's were not always part of the companies' vocabulary. In fact, until the late 1980s, companies wishing to go 'green', installed filters in the factories and then tried to get rid of the waste they had collected there. "People were just shifting their pollution from one place to another," states Ron Humber for Friends of the Earth in London. "What we need is a completely new approach to waste management."

"I'm dreaming of a factory," says Randy Newbury, "in which scrap vehicles enter through the back door for 100 per cent recycling and leave as new cars through the front door."

"At the moment this is only a dream," he quickly adds, "but we must do something now. Every year millions of cars in Europe go to the scrap dealers."

"In the past, industry was only able to recycle the metal parts. But after they had taken them out they still had up to 250 kg of mixed waste – plastics, glass, rubber and engine oil. The scrap dealers had to pay £50 a tonne to dump this waste – and the costs were rising. Now special firms collect and dismantle old vehicles and return as many parts as possible to the car manufacturers.

We code the plastic parts of each car so it is easier to separate material for recycling. In this way we reduce the amount of mixed waste, but there is still some left. So our engineers and suppliers are working on this problem to find substitute materials for the rest. Ideally car parts should be made from one family of material. All this adds to the cost of a new car but, of course, we are able to save money by re-using or recycling the parts.

This means that one day we will produce a car which is 100% recyclable, which allows us to use the same materials over and over again. As we see it every new vehicle we design in this way helps us to solve tomorrow's problems today."

493 words

a When you have a look at the picture and know the headline of the text "Vehicles go round again" – what do you expect the text to be about?

R **b** Skimming: → SF 2

Read the text quickly – as if you were reading a newspaper article – and say in short what the text is about.

R **c** Scanning: → SF 2

The headline above the photo is "The three R's of waste management".
Read the text again and try to find out what the three R's stand for.

d Find a suitable meaning of the following words taken from the text in the short excerpt from an English-English dictionary. Say why you have chosen the entry.

1. raw - line 13
2. shifting - line 24
3. dealers - line 35
4. rubber - line 39
5. dump - line 40
6. substitute - line 48

code *(n)* **1.** A **code** is a set of ideas by a group of people about the proper way to behave. **2.** A **code** is a set of written rules which state how people should behave in a particular country or society. **3.** A **code** is a system of changing a message by replacing all the letters with other letters so that the message is a secret. **4.** A **code** is a short group of numbers or letters which is used to identify something.

code *(v)* **1.** If you **code** a message, you replace all the letters with other letters so that the message is a secret. **2.** To **code** something means to identify it by a short group of numbers or letters.

dealer *(n)* **1.** A **dealer** is a person who buys and sells things. **2.** A **dealer** in a game of cards is the person who gives out the cards to the other players.

dismantle (v) If you **dismantle** a machine, you carefully take it to pieces.

dump *(n)* A **dump** is a place where rubbish is left.

dump *(v)* **1.** If you **dump** something, you throw it away carelessly. **2.** If a company **dumps** goods, it sells large quantities at very low prices, usually in another country.

raw *(adj)* **1.** Meat or other food that is **raw** is not cooked. **2.** Raw substances are in a natural state before being used in a manufacturing process.

recycle *(v)* If you **recycle** things that have already been used, such as bottles or paper, you process them so that people can use them again.

reduce *(v)* To **reduce** something means to make it smaller in number, size, price etc.

re-use *(v)* When you **re-use** something, you do not throw it away. You use it again.

rubber *(n)* **1.** **Rubber** is a strong, waterproof, elastic substance that is made from the sap of a tropical tree or is produced chemically. **2.** A **rubber** is a small piece of this material that you use **to rub** out mistakes that you have made while writing or drawing. **3.** A **rubber** is a series of games played between the same two people or teams.

scrap *(n)* **1.** A **scrap** of paper is a very small piece of it. Usually you do not want it any more. **2.** **Scrap** or **scrap** metal is metal from old or damaged machines, cars, etc. which is melted so that it can be used again.

shift *(n)* A **shift** is a change from one position to another.

shift *(v)* If you **shift** something, you move it or change its position.

substitute *(adj)* The word **substitute** can also be used as an adjective: "She had become attached to her **substitute** parents."

substitute *(n)* A **substitute** is a person or thing that takes the place of another.

substitute *(v)* If you **substitute** a thing, you put the new thing in the place where the old thing was.

R **e** Answer the following questions on the text.

1. What is the manager for environmental problems dreaming of?
2. How are other firms attacking the waste problem?
3. In what way did companies try to solve the problems of pollution in the 1980s?
4. How can special firms improve the work of the scrap dealers?
5. How can car companies help to solve problems of pollution when they are designing new cars?

1 Ecology or economics?

a Listen to the news report on Radio 4 and find the correct answers. R

1. The news report is about
   a. a North Sea oil project
   b. a zinc ore mining project
   c. a coal mining project
   d. an environmental project
2. The reporter mentions the effect of the project
   a. on the tourist industry
   b. on the air
   c. shipbuilding industry
   d. other mining companies
3. The people against the project are
   a. technologists
   b. workers
   c. environmentalists
   d. house owners

b Now listen to the recording again and find out more about the following points: R

1. where the company has found the zinc ore deposits
2. why this area is especially interesting for the company
3. how many jobs the project will create
4. what this development will probably mean for nature in this area
5. the fears of the hotel owners

c Use your answers to talk about the project. I

2 Zinc UK project: A public enquiry

Zinc UK, a mining company, have found large deposits of zinc ore in the attractive coastal area of Cornwall. The ore is rich in zinc and is near the surface. This means mining will be relatively easy and therefore profitable. The project will create new jobs for the area.
The problems are that this is a particularly beautiful area and it is the home to many wild plants and animals. It is also a tourist attraction for the visitors to the small seaside resorts along the coast. The seaside resorts are small because of the poor road communications and there is no railway. The mining company has promised to build a new major road which will go directly to the motorway 25 miles away. Unemployment in the area is high. Most of the jobs are on local farms and in the seaside resorts. However, in the last few years the seaside resorts have had fewer visitors. Local taxes for house owners are high and the local government wants to keep tax increases as low as possible in the future. Zinc UK would contribute to the income of the area. The small town of St. Agnes has a new housing estate, but it only has a few shops, a coffee bar, a pub and a primary school. There is no public library, sports centre or cinema in the area.

228 words

a What advantages/disadvantages do you think the project could bring to the area?

I **b** The local MP for North Cornwall has decided to hold a public enquiry to hear the views of the people concerned. The enquiry will help to decide whether Zinc UK should be allowed to start their mining project or not. The following people are at the meeting:

1. managers of the mining company
2. teenagers from the area
3. members of the local environmental group
4. hotel owners from the area
5. unemployed house owners from the area

Imagine you are one of these people.

1. List the advantages and disadvantages that the mining project might have for you personally.
2. Play your role at the meeting. Use expressions like "In my opinion" etc. to introduce your arguments.

P **c** After you have discussed the issue write a report of the meeting. Use reported speech. → SF 7+8

Unit 7

Energy

P **a** Copy the following mind map into your exercise book. Then complete the mind map with words of your own. → SF 9

b Compare your mind map with a partner (or in a group)
P and complete your notes.

c Choose one source of energy and present it in class with the help of your mind map.

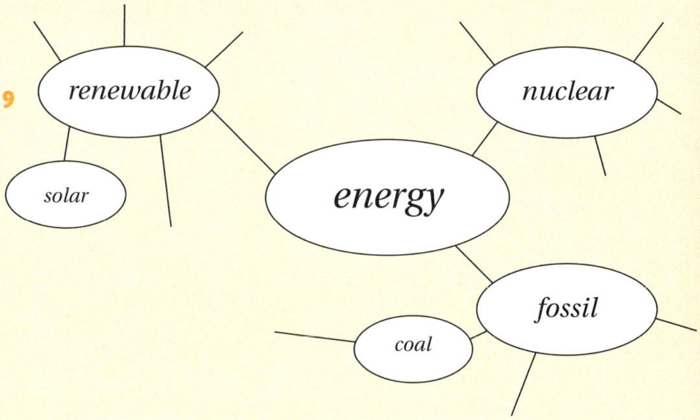

1 More wealth – less energy

Modern technology and relatively low prices have led to a situation in which most people feel relaxed about energy. Technology has made us believe that there will always be an answer to future problems and the prices still create the impression that we have plenty of energy. But we have little reason to feel secure.

More than half of the world's population (mostly in the developing countries) lives without a commercial supply of energy. The total energy consumption per head in countries such as India is only about 3 % of that in the United States. By 2100 the number of people in the world will rise to about 10 billion (compared with more than 6 billion today) and 85 % of mankind will live in poor countries. Either billions of people will continue to live in poverty, or the demand for energy will grow enormously. Governments and companies will invest trillions of dollars in the next twenty or thirty years – more than ever – in the developing countries. And if they do so, economic growth will mean much higher consumption of energy in the Third World.

Although the availability of fossil fuels is limited, experts predict that during this century the world will still depend on them, especially coal which is plentiful in India and China – countries with the highest populations. The problem is that developing countries care far more about the standard of living of their citizens today than they do about pollution and global warming. That is why they use the little money they have to increase the production of goods. If they had enough money, they would be able to produce and use energy in an environmentally-friendly way.

On the other hand many wealthy industrialized countries are introducing methods to save energy and they are further developing technologies which provide new sources of renewable energy such as solar power. Of course, it is not an easy job to transfer these expensive technologies to the poorer parts of the world. But if we help the developing countries, they will be able to increase their energy efficiency. In this way these countries will manage to raise the standard of living significantly. At the same time they will consume only little more energy than today and the production of this energy will not harm their environment so much.

The time to take action is now. The West is beginning to learn from the mistakes it has made. Surely it is important to help the developing countries so that they can avoid the same mistakes. If we do not act now, what will life be like in 2100?

435 words

R **a** Refer to the text and correct the following statements.

1. We must start to help the developing countries in the year 2100.
2. The number of people in the world is falling.
3. Today the United States consumes 5 times more energy per head than developing countries in South Asia.
4. In many developing countries people have developed methods to save energy.
5. The developing countries do not buy new energy technologies because they do not want to save energy.
6. The price of energy is very high today.

R **b** Find the nouns in the text.

1. grow 2. consume 3. pollute 4. supply
5. demand 6. produce 7. poor 8. efficient

c Find antonyms in the text. R

1. less 2. poor 3. high 4. a lot 5. past
6. industrialized (countries) 7. difficult
8. slighty 9. cheap 10. (the number will) fall

d Answer the following questions. R

1. Which are the sources of energy mentioned in the text?
2. Which of them is a fossil fuel and which is a renewable source of energy?
3. Which of the following are fossil fuels, which are renewable sources: gas, water, wind, oil, peat?
4. Which advantages and disadvantages do fossil fuels and renewable sources of energy have? Think about these points: geographical position, availability, costs, pollution.

2 Talking about graphs → SF 21

Graph 1: Development of world temperatures

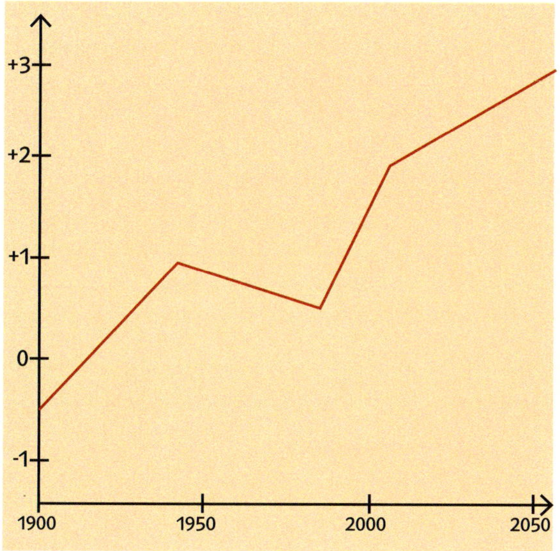

Graph 2: Development of energy consumption (1970 – 2020), in Quadrillion Btu (British thermal units)

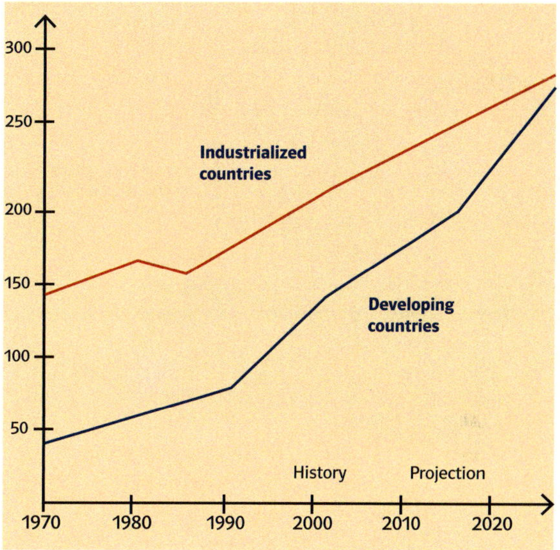

a Study the graph above and its description.

The graph shows the development of world temperatures.
Between 1900 and 1940 world temperatures rose significantly by 1-5 degrees.
From 1940 to 1980 they fell slightly.
Since 1980 temperatures have increased constantly.
If this trend continues, temperatures will be about 3.5 degrees higher in 2050 than in 1900.

b Now describe graph 2. Use the words in the boxes below.

Trends	
increase increase rise go up grow	**decrease** decrease fall go down drop
remain stable steady constant at the same level	

Changes	
big changes significant(ly) substantial(ly)	**small changes** slight(ly) a little
fast changes sudden(ly) sharp(ly)	**slow changes** slow(ly) gradual(ly)

1 How to save energy in your house

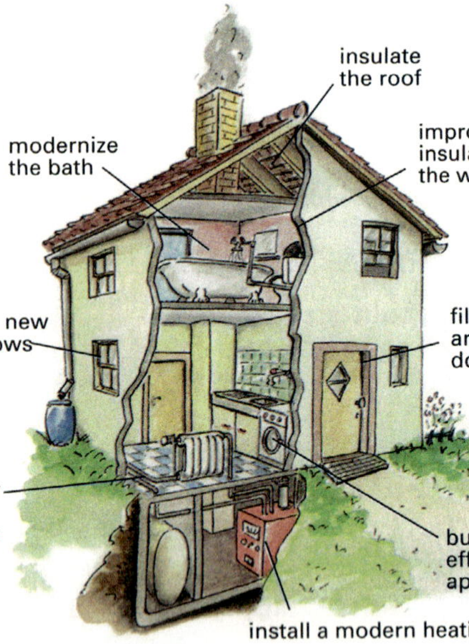

Darren and Kate Palmer have just bought a house. The house is 25 years old and needs some repairs. The Energy Council has sent them a brochure on ways they can save energy in their house. Darren is showing Kate the brochure and says: "Look, it says here …"

Example:

"If we insulate our roof well, we will save a lot of energy."

■ **Use the drawing and the phrases in the box to make at least eight sentences in the same way.**

lose less heat • save money • not use so much hot water • avoid a loss of heat • reduce our energy consumption • not pollute the environment so much • cut our water bill • not waste so much energy • save a lot of energy

2 The advantages of alternative energy

Cathy Johnson, a politician and expert on energy questions, is giving a talk at a local school. She is convinced that we depend too much on the use of fossil fuels (oil, gas and coal) as sources of energy. She believes that we should save energy and also look into the possibilities of the use of alternative energies more carefully. Here are some of her notes for her talk.

■ Use the expressions in the box to express her opinions.

1. First of all – must say
2. Secondly – be of the opinion
3. Thirdly – think
4. At this point – must emphasize
5. Moreover – must add
6. In addition – believe
7. Furthermore – must say
8. Finally – be convinced

Example:

1. First of all I must say, if people used alternative forms of energy, they would contribute to …

1. **people use alternative forms of energy – they contribute to a cleaner world**
2. government spend more money on these alternatives – we all see the advantages in a few years
3. electricity companies replace old power stations in this area – they can produce electricity and not pollute the environment so much
4. firms produce more energy efficient appliances for the home – we not need so many power stations
5. house owners insulate their houses better – they can reduce their electricity bills
6. we use the sources of renewable energy in this part of the country – we save a lot of fossil fuels
7. people not travel by car so much – we not have such a big energy problem
8. we not waste energy as we do now – we have a cleaner and better world

3 Bob Flynn is thinking about his life

Bob Flynn is the manager of a small company which develops solar systems.

■ Express his thoughts for him like this:

Example:
If I had worked hard, I would have got good marks in the first year. If I hadn't spent more time on my studies, I wouldn't have done well at college in the end.

1. I got a job as an electrician – after I wrote a lot of letters of application.
2. I went to university – after I attended evening classes three times a week.
3. I studied electrical engineering – after I looked at the good job possibilities.
4. I passed my examinations at university – because I worked hard.
5. I didn't go straight into my own business – because I wasn't sure what to do after university.
6. I started up my own business – after a friend gave me some tips.
7. I bought a workshop – after the bank lent me $50,000.
8. I became interested in alternative forms of energy – after I joined an environmentalist group.
9. I started to develop solar systems – because I learnt all about solar energy at university.
10. I didn't invest more money in the business – because I didn't have enough money at that time.

4 Save energy now

Two students have started a "Save Energy Now" group. They are talking about the use of energy and how people have reacted to new developments.

■ Put the verbs in brackets into the correct form.

Mary Jo: … but we still aren't doing enough and if we **1. (continue)** to take so little care of our environment, soon we **2. (not have)** an environment at all.
John: I know. If people **3. (know)** years ago about the damage that power stations can do to the environment, I'm sure they **4. (do)** something about it a long time ago.
Mary Jo: Yes, but now we all know about it and we can all help. If a lot of people **5. (not be)** as lazy as they are, they **6. (not waste)** so much energy in their homes. It's the little things that help.
John: Yes, you're right. I'm sure that if the school **7. (not help)** us with our project, then we **8. (never be able to start)** it, for example.
Mary Jo: That's true, and nowadays there are more people who are really interested in alternative forms of energy. If that **9. (not be)** so, we **10. (not find)** so many people at our meetings.

(Heading?)

(A) In the mid-1970s fewer than 1,000 U.S. homes generated their own power. That number has grown to more than 100,000 and every year thousands more will join those who are already off-the-grid. They are turning to renewable energies, such as wind, water and sunshine in order to generate power for their homes. But why are more and more people doing this? Very often it is because these people live far away from the grid and cannot afford the cost of a power line to their home.

(B) As a result of this strong interest, it has become big business to supply homes with renewable energy. The yearly sales for the solar-energy industry alone have already reached $1 billion, and wind and hydro systems for domestic users add millions more.

(C) Wind power is generally used for industrial and commercial applications. However, the most popular off-the-grid technology for private homes is the use of solar energy. Although photovoltaic (PV) systems are nearly twice as expensive as wind power, the sun provides energy for more than two thirds of America's off-the-grid homes. This is because solar energy is easy to install and doesn't need a lot of maintenance. In the past, makers of PV systems sold 60 % of their equipment to the government.

(D) However, their main customers now are private homes and companies who have realized that PV systems are often cheaper in the end. For example, private companies can install a PV cell on phones or on highway signs which are far away from the grid and so it's not necessary to dig up roads and lay cables. Even the building industry has shown an interest, offering solar-equipped homes at no extra cost.

(E) One of the most important developments recently is that some electricity companies are even using solar power to avoid the need to build more traditional power plants. A Californian gas and electricity company has supplied 150 homes in California with electricity from a number of PV panels. In this way it saved the $ 1 million it would have spent on a traditional system.

(F) "All this is going in the right direction," says Tom Kelly, a solar research scientist, "but we must still make a number of engineering improvements before we are ready for the mass market." In fact, research is going on all the time in order to find cheaper and better materials and to increase efficiency in solar energy installations. Off-the-grid energy may even get some help from politicians, who may introduce an extra tax on fossil fuels in order to limit global warming. If they do this, fossil fuel prices will increase and solar energy will therefore become more attractive.

(G) However, the energy of the future might come from a completely different source. Many experts believe that hydrogen-based fuel cell systems will provide the energy for the generations to come. Fuel cell powered cars are already on the roads and many companies are testing fuel cell systems. They are relatively cheap, renewable and environmentally-friendly. So, let's see what the future holds for us!

536 words

R **a** Read the text quickly in order to find a suitable heading. → SF 2

R **b** Match the following headings with a suitable paragraph. There are two headings that do not fit.

1. The economic success of renewable energy
2. Solar systems need to be improved
3. Fuel-cells the future answer to energy problems
4. Solar systems more expensive than wind power
5. Private customers and companies especially interested
6. Solar energy most popular
7. Growing interest in renewable energy
8. Solar power stations replace traditional power systems
9. Off-the grid homes in the USA

R **c** Find the most suitable option.

1. Many people in the US generate their own power because
   a) they do not want a power line.
   b) a power line to their homes costs too much.
   c) they prefer renewable energies.
   d) they hope it is big business.
2. The government
   a) is no longer the biggest customer for PV systems.
   b) buys three out of five PV systems.
   c) now buys more PV systems than private customers.
   d) was the second biggest customer in the past.
3. In California a gas and electricity company
   a) is building a new traditional power plant.
   b) has spent $1 million on a PV system.
   c) has supplied 150 homes with PV systems.
   d) is using energy from alternative sources to supply homes on the grid.
4. There are several advantages of hydrogen-based fuel cell systems given in the text. Which one is not mentioned? This source of energy …
   a) is unlimited.
   b) is relatively cheap.
   c) is environmentally-friendly
   d) will mean big business.

M **d** Mediation → SF 10

1. Erläutern Sie die folgenden Ausdrücke aus dem Text kurz auf Deutsch.
   a) "those who are already off-the-grid" (lines 5/6)
   b) "wind and hydro system for domestic users" (lines 19/20)
   c) "fuel cell powered cars" (lines 71/72)

2. Übertragen Sie die Abschnitte F – G ins Deutsche. Übersetzen Sie nicht Wort für Wort und schreiben Sie ganze Sätze.

P **e** Write a summary of the second part of the text (E – G). It should be about half the length of the original text and you should use your own words. → SF 6

1 North Sea oil output falls

R **a** Listen to the news report on Radio 4 and note the nine figures that are mentioned.

R **b** Now listen to the report again and find further information about each figure.

P **c** Prepare a short report about the development of oil production and consumption with the help of your notes. → SF 7+8

2 Project work → SF 13, 14, 22

This is an island off the west coast of America. The island has strong winds from the west, high mountains with fast running rivers, thick forests along the west coast and peat bogs in the south east. There are no resources of coal, oil or natural gas on or near the island. At present only a few people live there.

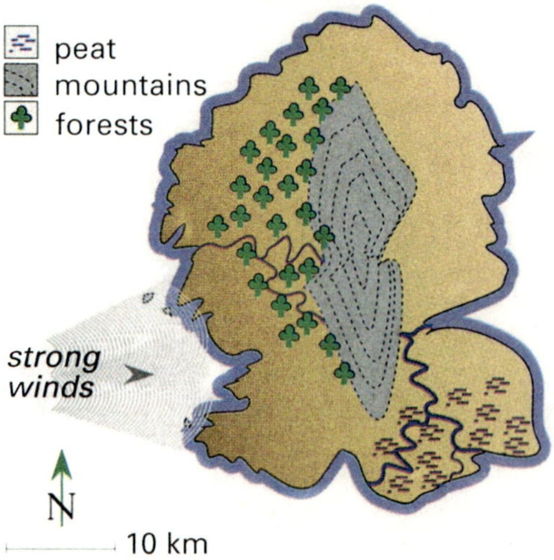

You are in a project team which wants to develop the island. P

Think about and discuss these ideas.

1. What material would you use to build homes on the island?
2. Which two materials could help you to produce heat to cook food?
3. What could you use to produce electricity to run machinery etc.?
4. Where would you build houses on the island? Give reasons.
5. What environmental problems will you cause when you use the natural resources on the island?
6. What conclusion do you draw from your study?
7. Present the result of your project team in class.

Down Under

a What do you think does 'down under' mean?

b Quiz on Australia. If necessary use the Internet to find the correct answer.

1. It takes about … ? … hours to travel to Australia.
   a) 3
   b) 9
   c) 24
   d) 48

2. If you travel from Munich to Sydney, your journey will be about … ? … km long.
   a) 1,000
   b) 7,000
   c) 14,000
   d) 30,000

3. The capital of Australia is
   a) Perth?
   b) Melbourne?
   c) Canberra?
   d) Sydney?

4. Australia is …
   a) as large as Germany?
   b) twice as large as Germany?
   c) ten times larger than Germany?
   d) twenty times larger than Germany?

5. Compared to the population of Germany the population of Australia is
   a) a tenth?
   b) a fourth?
   c) the same?
   d) twice as much?

6. The native Australians are called
   a) Originals?
   b) Origins?
   c) Aborigines?
   d) Unoriginals?

7. Head of State is the
   a) Prime Minister?
   b) King/Queen of Australia?
   c) President?
   d) Chancellor?

Quiz on Australia

1 Travelling in Australia

Location: http://www.backpackingaround.com

About us | About Australia | Employment & Accommodation | Notice Board | Travel Tips | Pubs, Clubs, Cafes & Shops | Activities

Welcome to Backpacking Around

Hi I'm Bruce.
I'm here to help you on your backpacking trip to Australia.

(A) It is best to contact your nearest Australian embassy or consulate to find out what visas you will need (some travel agents may also be able to assist). As a general rule - you will need a valid passport and visa in order to travel to Australia. Visas are free and are for a 3 month period. The tourist visa does not allow you to work or study in Australia. And these days, it means exactly that. As employers require a tax-file number from all employees, you really can't work without the correct visa. If you wish to work or study, enquire about obtaining a working or student visa.

(B) The essential components of good travel insurance fall into three groups – medical insurance, cover for your luggage and personal property, and cancellation insurance. If you are planning to have a bit of adventure on your travels, it is wise to check if your insurance policy covers activities like parasailing or motorbike riding.

(C) Check your baggage limit (generally 2 pieces of checked luggage to a maximum of 20 kg for economy passengers). Remember, the safest weight for a backpack is a third of your own body weight. Try and keep your luggage to a minimum – if you aren't 100% sure you need it – leave it out. You can always buy it in Australia (but chances are you won't miss it!). If travelling in Australia's winter, don't forget a warm jumper or two! "Roll" don't fold - it takes up less room and makes less wrinkles!

(D) If travelling with a friend, pack half your clothes in their bag, and carry half theirs in your luggage. Take at least two different kinds of "money", traveller's checks and a credit card for example, and put them in different places. Always carry all valuables, cameras, money, traveller's checks, keys, travel documents, and a change of clothes with your carry-on-luggage. Clearly label all luggage, with your name, address, and phone number. Always remember to pack your own luggage and lock it securely. Never leave your luggage unattended. Never carry any parcels for other people.

(E) Calling home from Australia is relatively inexpensive. Probably the cheapest option is to purchase a calling card which is widely available. Calls made using such cards generally cost around 5 cents a minute. Before you leave home set up a free email service from one of the many sites that offer them. You will be able to access the Internet from all around Australia, either from your hostel or from an Internet café. Generally, you will only have to pay a few Australian dollars for 30 minutes access. Sending postcards may seem a little old-fashioned but is still one of the best ways of contacting home! Postage stamps can be purchased at Australia Post shops in most major towns and even at some news agencies and small shops.

500 words

R **a** Read the text on page 66 quickly and match five of the following headings with the corresponding paragraphs. Two headlines do not fit. → SF 2

1. Safety tips
2. Internet links
3. Entry regulations
4. Keeping in touch
5. Valuables
6. Perfect packing
7. Useful insurances

b Now study the travelling tips more carefully and complete the sentences.

1. When you want to travel to Australia you must have a … and a … .
2. As a tourist you can't … or … .
3. Instead of cash you can take … … or a … … with you.
4. Telephoning home is not too expensive with a … … .

R **c** What can happen if you …

1. … leave your bags unattended?
2. … carry parcels for other tourists?
3. … go over the limit of 20 kg?
4. … do not label your luggage?

d How can it help if you … R

1. … use a free email service?
2. … have a working visa?
3. … have a cancellation insurance?
4. … keep money in different places?

2 An email to a friend

P ■ Imagine you are in Australia. You are now in Sydney again at the end of your stay in Australia. Write an email to an English backpacker friend you met in Sydney at the beginning of your trip. Describe your journey and what you have seen and done so far. → SF 7+8

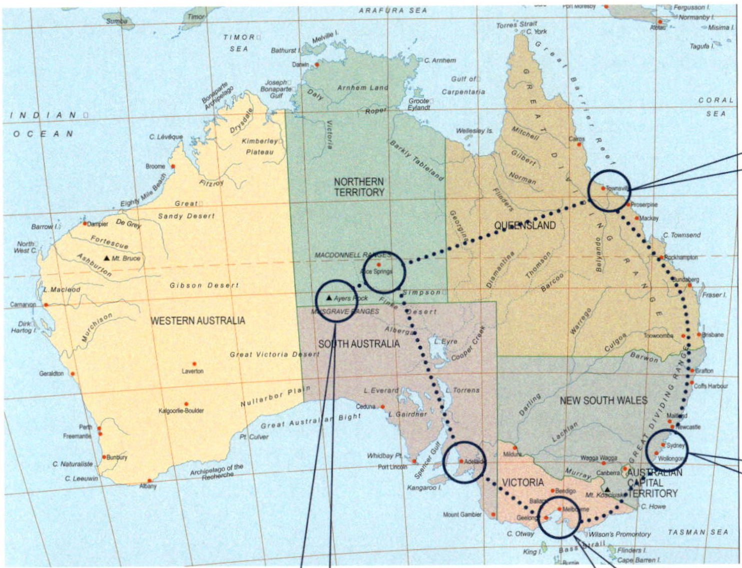

1 How a website is created

a This is the official site of the Australian Tourist Commission. With the help of the vocabularly below describe how such a website is made.

Example:

First an Internet address is invented.
Then photos are….

1. an internet address - invent
2. photos - take
3. text -write
4. layout - design
5. corrections - make
6. links - add
7. other internet addresses - check
8. all the information - store
9. website - register

b How can the website for backpacking around (see page 66) be improved ?

In order to have a successful website it has to be made attractive. How can this be done ? Here are some suggestions which backpackers to Australia have made. Complete the sentences in the passive voice. Must, should, can or could should be used at least once.

1. more photos of Australia / show
2. the cartoon figures / take out
3. the text / write in larger letters
4. the paragraphs / give headlines
5. more special tips / recommend to the reader
6. more internet addresses which you can contact / introduce
7. more interactivity / arrange
8. these suggestions / follow up / by the site managers

2 Australia's early history

The first people who discovered Australia were hunter-gatherers who arrived via land bridges and short sea-crossings from Asia. European settlers have referred to these people as Aborigines. In 1770, James Cook mapped the east coast of Australia, which he named New South Wales and claimed for Britain. The British government then established a penal colony in Australia, following the loss of the American colonies that had previously filled that role. Captain Arthur Phillip founded the first colony (Port Jackson) in south-east Australia on 26th January 1788. Since the 19th century, Australians have celebrated this date as Australia Day, the country's national holiday. Many towns and cities around the country hold fireworks displays and other celebrations on this day. For many years, almost half the population of the city of Perth have watched Skyworks, the largest Australia Day celebration.

■ Use the passive voice and form sentences about the early history of Australia.

Example:
Australia was discovered by hunter-gatherers who arrived via land bridges and short sea-crossings from Asia.

3 A gap year in Australia

After graduating from secondary school, Susanne Meier took part in a gap year programme in Australia, which included working on a ranch in the outback. Here are some of her experiences:

1. On my arrival / Andrew, a member of staff, / meet /at Brisbane Airport
2. Afterwards / take / to Brisbane backpackers hostel / where I spent the night
3. On the following day / picked up / Mrs. Anderson, my boss's wife, / drive to their ranch
4. Over the next four days / show / how to work with cattle and horses
5. In addition / the Andersons / teach / about fencing and chainsaw work
6. I learnt the basics necessary for the jobs that / offer / the farm
7. For the duration of my stay / two horses / give / which I had to shoe myself
8. Sometimes / allowed / to sleep outside under the stars, which was an awesome experience
9. Even though / expect / to work hard in extreme heat / it was an unforgettable experience

■ Form sentences both in the active and passive voice.

Example:
1. On my arrival, I was met by Andrew, a member of staff, at Brisbane Airport.
On my arrival, Andrew, a member of staff, met me at Brisbane Airport.

The Australian Tourism Industry

1 Describing a picture → SF 19

a Describe what you can see in the picture. P

b Answer the following questions. P

1. What ideas come into your head when you read the caption?
2. What do you think the visual elements represent?
3. What do you think is the message of this picture?
4. Provided that this picture is part of a promotion brochure for Australia, would it persuade you to go to Australia? Why/why not?

2 Describing a cartoon → SF 20

P **a** Describe and analyse the cartoon on the right.

1. The illustration shows a kangaroo mother and …
2. The baby kangaroo …
3. Its mother is wearing …
4. The cartoonist is making fun of a typical situation …
5. I think …

P **b** Describe and analyze the following cartoon in a similar way.

We can talk now: Mum has reluctantly agreed to give me some privacy

"It looked a lot bigger in the brochure."

3 Describing and analyzing diagrams → SF 21

a Study the following pie chart and complete the text below.

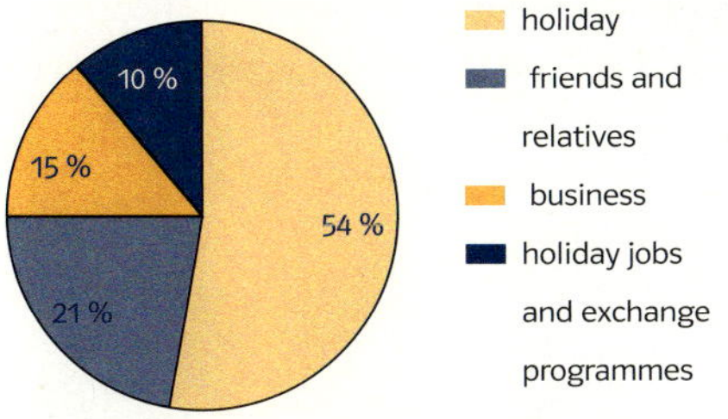

The main purpose of travelling to Australia
During the year 2003 there were 4.4million visitors to Australia. More than half **(… 1 … per cent)** were holiday visitors, nearly a quarter **(… 2 … per cent)** were visiting friends and relatives and **(… 3 … per cent)** were business visitors. One in ten **(… 4 … per cent)** went there to do holiday jobs or to take part in student exchange programmes.
To conclude you can say that three quarters **(… 5 … per cent)** of visitors travel to Australia for pleasure.

P **b** Describe and analyze the following bar chart. → SF 21

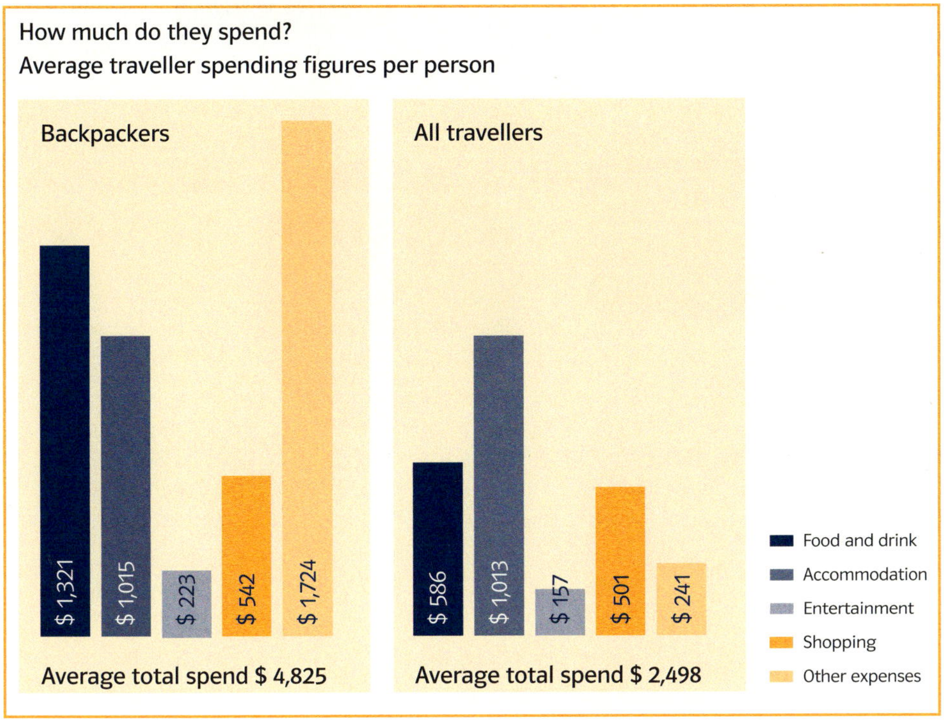

1 Voting systems – Great Britain – Denmark – Australia

R **a** First listen to the three politicians who are talking about their own political careers and find out about the following points:

1. Name
2. Country
3. Political party
4. Special interests
5. Constituency
6. Former job
7. Member of Parliament since

P **b** Now describe the three politicians in complete sentences.

R **c** Listen to the three politicians again. Try to find out about their voting systems.

P **d** The first politician went on to talk about the following statistics. What do you think he said?

United Kingdom General Election 2005

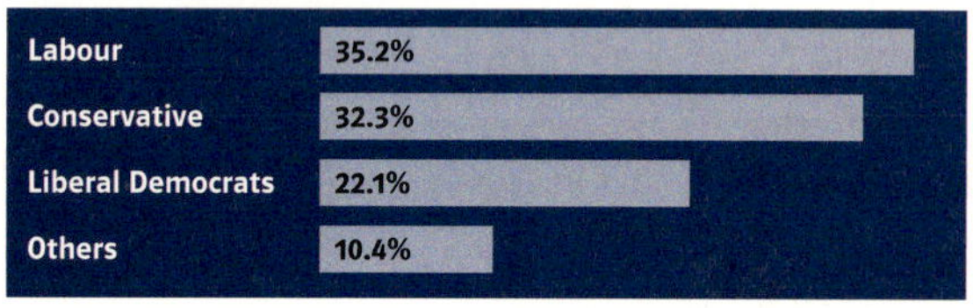

Distribution of seats

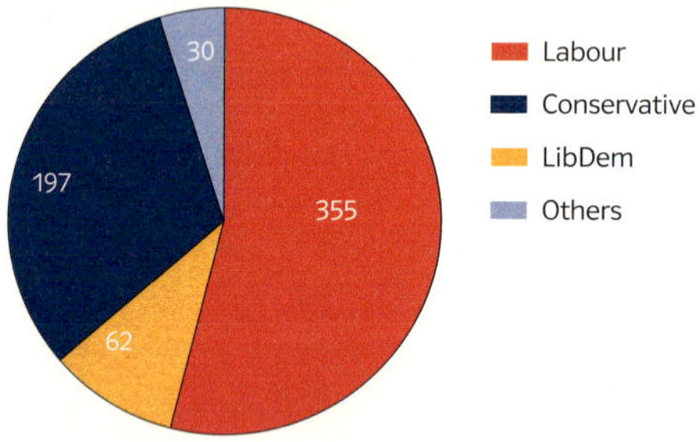

2 The Australian parliamentary system

P **a** Describe the diagram of the Australian parliamentary system.

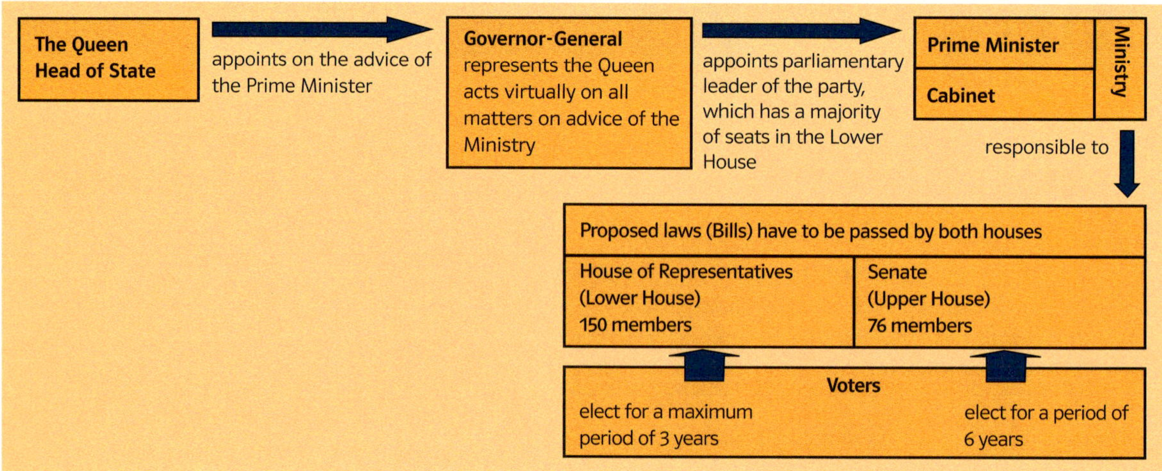

P **b** Work with a partner or in small groups and prepare a poster either about the American or British parliamentary system. Use information from the internet.

Young people in society

a Before you listen to the song on the right, what do you think the song is about?

R **b** Now listen to the song and answer the following questions.

1. What do you now think the song is about?
2. Who or what do you think the singer is looking for?
3. Who do you think is "you" in the song?
4. What has the song got to do with you?

I still haven't found what I'm looking for (U2)

I have climbed highest mountain
I have run through the fields
Only to be with you
Only to be with you

I have run I have crawled
I have scaled these city walls
Only to be with you
But I still haven't found
What I'm looking for

I have kissed honey lips
Felt the healing in her fingertips
It burned like fire
This burning desire
I have spoke with the tongue of angels
I have held the hand of a devil
It was warm in the night
I was cold as a stone
But I still haven't found
What I'm looking for

I believe in the Kingdom Come
Then all the colours will bleed into one
But yes I'm still running
You broke the bonds
You loosed the chains
Carried the cross
And my shame
And my shame
You know I believed it
But I still haven't found
What I'm looking for

The missing

Every year around 100,000 children run away. Many are escaping abuse or trouble at home, but some just go.

Sarah Turner held a poster of her missing 14-year-old daughter up to the face of a teenage girl of similar age, living on the streets of South London. The girl, a crack addict, read how Ruth, a bright and attractive schoolgirl, had disappeared outside her local library – leaving her parents unable to imagine why she should do so. As the girl read, tears rolled down her cheeks. Then she shook her head and said that she was sorry but she had not seen Ruth. "I'm sure she cried because she recognised so much of her own story in Ruth's," says Sarah Turner.

Until Ruth disappeared Mrs Turner had been unfamiliar with the world of life on the streets. But after Ruth went missing last month, the police reported a sighting of her in the central London area. So Mrs Turner began to search every street and every corner. She became well-acquainted with the area's outdoor population. "The police told me that looking for runaways like Ruth is like looking for a needle in a haystack," she says. "I knew that if I wanted her back, I would have to find her myself and find her quickly before something terrible happened to her." She searched day and night, while her husband Paul and her elder daughter Joanna waited by the phone at home.

This was not the first time Ruth had run away, but it was the longest she had ever been missing. For a few years, she had left home from time to time, first only for a few hours, later staying away overnight. Although she often stayed away from the exclusive girls' school she attended, Ruth seemed able to keep up with her work. "All she said, each time she returned," says Sarah Turner, "was that she had sat on a bus and waited to see where she would end up. It didn't matter to her."

The lack of explanation for her flights has been one of the hardest things for her parents. There has not been a single accusation in a moment of anger. All her mother knows for certain is that Ruth suffers from low self-esteem and seems unhappy at school. "She doesn't have a lot of friends and I think she feels isolated. She thinks differently to other kids, and from a young age, she was always asking questions which have no answers like: *Why did the universe start?* Maybe she's frightened of something."

Ruth is one of about 100,000 under-16s who go missing every year. One in three boys and one in five girls are under 11 when they first run away. From the age of 14 upwards, girls are twice as likely to run away as boys. 45% of children in care run away compared with 9.5% of those who live with their families.

Happily, Ruth turned up safe and well after 10 days away. When the police collected her from a homeless project where she had sought help, the officers said that they were convinced that she had not been abused in any way and that she had not taken drugs. But Ruth has refused to see her family and chosen instead to go into foster care. Social workers have told her family that they cannot see her at present. For her parents this is a painful time as they wait for social services to permit a meeting, for her mother in particular. She says: "I just miss her so much and want her to come home again."

590 words

(The Guardian, 31 October, 2001, adapted)

R **a** Match the beginnings of the sentences with the correct endings. Note: Two endings are wrong.

1. Sarah searches the streets of London
2. The crack addict cried
3. Ruth had run away several times before
4. Ruth seems unhappy at school
5. Although Ruth was found safe and sound

A. because she can't find any real friends there.
B. she doesn't want to go back to her family.
C. because her parents treated her badly.
D. because her daughter had run away from home.
E. because she saw herself in Ruth's story.
F. because she was hungry.
G. but her parents never found out the reason.

R **b** Answer the following questions:

1. Why did Sarah Turner search the streets of London?
2. What was Mrs Turner afraid of?
3. Why may Ruth have left home?
4. What children are most likely to run away?
5. Why – do you think – did Ruth not want to return?

c Complete the following sentences with words from the text on page 74.

self-esteem • homeless • disappeared • search • reappeared • missing • in foster care • isolated • library • overnight

In the text "The missing" the author describes a mother's **(1)** … for her missing teenage daughter. Day and night Sarah Turner walked the streets of London carrying a photo of her **(2)** … daughter, Ruth. She asked all of the people on the streets, especially in the area of South London where her daughter had been seen last. Ruth had **(3)** … without a trace near her local **(4)** …. Nobody really knew why she had left. Sarah met many **(5)** … people and she began to realize how difficult it would be to find her daughter. She also spoke to another teenage girl who reminded her of her own daughter and the dangers of life on the streets. Ruth had gone missing quite a few times before, but had never been away from home for such a long time. The first couple of times she only left home for a few hours, later she stayed away **(6)** … . She did not go to school regularly and when she came back, she did not know why she had left. She seemed unhappy at school, had low **(7)** … and did not have many friends. Her mother believed that Ruth felt **(8)** … and often did not receive appropriate answers to her questions. According to the text, about 100,000 children under 16 run away every year. Most of the older ones are girls, and children in **(9)** … often run away from home.
Ruth finally **(10)** … at a homeless project after ten days and was collected by the police. She was safe but did not want to return to her family. Ruth went into foster care leaving her parents to wait for her.

d Vocabulary

1. Look in the text to find synonyms for these words:
   a) run away
   b) problems
   c) intelligent
   d) good-looking
   e) go to (school)
   f) come back
   g) 20 percent
   h) two times
   i) sure
2. Explain the following:
   a) abuse (line 2)
   b) live on the street (line 6)
   c) addict (line 7)
   d) runaway(s) (line 23)
   e) like looking for a needle in a haystack (line 24)
   f) go missing (lines 52/53)
   g) homeless (line 61)
   h) foster care (line 65/66)
   i) social worker (line 66)

M **e** Übertragen Sie die Zeilen 1 bis 21 ins Deutsche. Übersetzen Sie nicht Wort für Wort. → SF 10

1 A Canadian student at a British college

Craig Russell, a Canadian student from Vancouver, is planning to go on a year's exchange visit to Newcastle College in the north of England. He has never been to Europe before. So yesterday he asked Calvin, another student who was on last year's exchange, for some help. Here are some of Craig's questions and the answers he received.

a Match the answers to the questions.

1. Who should I contact first at the college?
2. When should I write to my host family?
3. How should I travel from London Airport to Newcastle?
4. Where can I reserve a seat for that journey?
5. What should I give my host family as a present?
6. How much luggage should I take?
7. What should I wear at college?
8. Where can I find a good bank in Newcastle?

a) at a travel agent's
b) a month before you arrive
c) opposite the college
d) Ms Keegan, the head of department
e) not too much
f) by coach
g) casual clothes
h) something typically Canadian

b Now report Calvin's answers like this:

Example:
Calvin told him / advised him to contact Ms Keegan at the college.

c Craig has been at Newcastle College for one month now. This is part of Craig's letter home. As you can read the teachers **expect him to work** hard during his stay in Newcastle and sometimes they **make him do** some extra work. Complete the letter with the verbs in the box.

let • expect • make • want • ask

Life at college is great. I am the second Canadian student to study here for a whole year. I get on well with the teachers here but they **(1)** me work quite hard. They **(2)** me to talk to the English students more about the Canadian college system. I didn't realize it was so complicated. In fact, the teachers **(3)** me to know a lot more than I actually do. Last week, for example, they **(4)** to show my photos of our college in Vancouver in order to give the English students a better idea of our work at Canadian colleges. But it is not all hard work. And if I want to do some sight-seeing at the weekend, the teachers **(5)** me go early on Fridays. By the way I have met …

2 Finding out useful information

a On the next page you can see parts of the students' Internet notice board at Newcastle College. Match the following sentences with the notices on the Internet notice board.

Example: Sentence 1 goes with notice D.

see Grammar: Infinitive / Gerund → pages 185-186

b Complete the sentences below. Use the -ing form of a suitable verb.

Example: I'm interested in knowing more about France so I'll go to the French Club.

1. I'm interested in … more about France.
2. I'm tired of … with my parents and I need somewhere to live.
3. Students who are afraid of … their final tests should ring 347168.
4. … is dangerous for your health.
5. Craig is going to the Sports Club because he is good at … basketball.
6. We'll leave the car at home … and … is stupid.
7. I'm looking forward to … you on Friday evening.
8. If Gary takes me in his car, I won't have any more problems in … to college on time.

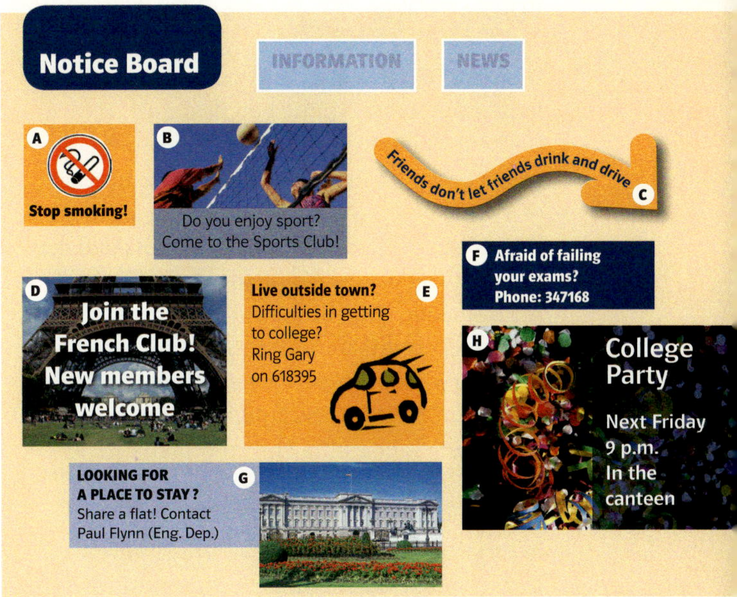

3 Talking about friends

Craig has met a lot of new friends at Newcastle College. Here are some remarks they have made about themselves.

Sharon: "I play more hockey than volleyball at the Sports Club."
Kevin: "I go to the cinema every week."
Emma: "I don't smoke any more."
Bill and Ben: "We don't eat in the college canteen. The food's not very good there."
Sam: "That's one thing I hate – my alarm clock."
Mary Ann: "I'm very interested in foreign languages."
Gary and Sue: "We did the same test last year."
Lisa: "I try not to be late for college. I take the early bus."
Mike: "I spend all my free time at the Sports Club."
Tina: "This book is very useful. I think you should read it."

■ Make statements about Craig's friends.
The verbs in the box will help you.

like • hate • prefer • remember • stop • enjoy • avoid • suggest • dislike • love

Example: Sharon prefers playing hockey to volleyball at the Sports Club.

4 Mediation → SF 10

Imagine you were at Newcastle College on an exchange visit and that you had to write a report in English about your experiences there.

■ Translate these notes into English.

1. Meine Gastfamilie ist sehr nett. Ich wohne gern bei ihr.
2. Aber zuerst hatte ich Schwierigkeiten, ihr Englisch zu verstehen.
3. Am Anfang hatte ich Angst, Fehler zu machen.
4. Die englischen Schüler im College sind sehr interessiert, mehr über Deutschland zu hören.
5. Im College lassen die Lehrer mich ziemlich hart arbeiten.
6. Sie erwarten, dass ich viel über das deutsche Schulsystem weiß.

Ecstasy – The dangerous "high"

It is midnight at a club in London. Lights flash, loud music plays, young people dance wildly to the beat. Outside there is a poster banning the use of Ecstasy. But few seem to notice it. Most of the teenagers inside the club have taken the drug, despite all the warnings about its dangers. For many teenagers, going to a club or a rave means taking "E". Those who use it say it makes them feel optimistic and warm about everything and everyone they see. One teenager describes it like this: "You don't feel cold, you don't feel pain. Everything is beautiful and you can dance all night."

Ecstasy is taken by about 5 million young people in Britain. The so-called "love-drug" started off in the United States, where it was often used by rich Yuppies as an expensive way of getting high. Nowadays Ecstasy is Europe's most popular drug for the 15 – 25 age group. Like all drugs, Ecstasy can be very dangerous. Over the last five years over 150 people have died after taking the drug. Death is caused when organs fail due to dehydration. Ecstasy raises the body temperature, and hours of wild dancing without enough to drink make it even more dangerous. The drug is often mixed with other ingredients in tablets: heroin, aspirin and dog-worm tablets have all been found in Ecstasy tablets which leads to high risks when taking the drug. Moreover, Ecstasy has long-term negative effects. Doctors believe it causes serious depressions in later life and may lead to permanent brain damage.

Most young people know about the risks involved. So why do they still take Ecstasy? Like most things that are dangerous, but also pleasurable – alcohol and smoking included – people tend to ignore the risks and concentrate on the fun. Furthermore, Ecstasy is regarded as an acceptable, even glamorous drug, unlike hard drugs such as heroin and cocaine. Drug counsellors, however, see nothing glamorous about "E". They see it as a dangerous way of trying to escape from life's problems. Often it is taken by unemployed teenagers as an escape from poor housing and the hopelessness of not being able to find a job.

The case of Leah Betts, an 18-year-old British girl, who died after taking Ecstasy, shocked many people in Britain this year. Leah was a typical "nice" young girl from a good home, who knew all about the risks involved. This case shows us all how difficult it is to convince young people that Ecstasy is very dangerous. Many parents will now worry even more when their teenage children go out to parties and clubs. Others will hope that the shock of this latest Ecstasy death will be enough to persuade their kids that it is not worth risking death for a few hours of feeling good.

468 words

R **a** After you have read the text carefully, answer the following questions in your own words as far as possible.

1. Why do many teenagers take Ecstasy?
2. Why have some people died after taking the drug?
3. What other negative effects can the drug have?
4. Why do people take the drug although they know about the dangers?
5. How do drug counsellors see the situation?
6. What does the death of Leah Betts demonstrate?

M **b** Explain the following words from the text in English.

1. Ecstasy **(headline)**
2. dehydration **(line 23)**
3. depression **(line 31)**
4. counsellor **(line 40)**

M **c** Übertragen Sie die Zeilen 33–57 ins Deutsche. Übersetzen Sie nicht Wort für Wort.
→ SF 10

P **d** Using your own knowledge and ideas, write a comment on the following statement: "The reports about the deaths of teenagers like Leah Betts will persuade young people not to take drugs like Ecstasy." → SF 5

P **e** Imagine you are a social worker at college. You have to write a report about your work this year. Use the following notes after you have put them in a logical order: → SF 7+8

- met the students outside the college (youth clubs etc)
- talked to their parents
- introduced myself to the students
- had individual talks with students who wanted to talk to me
- told about the dangers of drug abuse in general
- talked to the teachers
- held group sessions

P **f** Write a short text for an information leaflet on the risks of taking Ecstasy. → SF 7+8

g Complete the following text with words in the box.

people • play hard • studies • top • habits • drug use • distinction • drug active • figure • criminals • compared • legal • drinkers

Drinking and taking drugs is part of everyday life for young people in Britain. **(1)** … have found that British teens work hard at school, university and in their jobs, and **(2)** … in the evening and at weekends. In addition, young people make little **(3)** … between soft and hard drugs, from tobacco to alcohol to cocaine. 82% of people asked said they were weekly **(4)** … at the age of 18. Five years later, their drinking **(5)** … had not changed. At 18, 63% had tried illegal drugs, five years later this **(6)** … rose to 76%. Professor Parker of the Manchester University said: "Even now, more than half remain **(7)** … , with 30% being regular drug users, primarily of cannabis, followed by ecstasy and cocaine." Drug users have often been described as unemployed layabouts or **(8)** … . But these **(9)** … are young, educated and employed. It seems **(10)** … is part of normal life in Britain. Others surveys found that Britain was **(11)** … for ecstasy with 12% of young people saying they had tried the drug, **(12)** … to 3.7% in Germany. However, medical experts warn that even soft drugs (tobacco, alcohol or cannabis) can cause severe psychological and physical problems, although these substances are **(13)** … in many countries.

Social work with young people

a Listen and find out the following things: R
1. Who is talking?
2. Where is he talking?
3. What is he talking about in general?

b Listen and find out if the following statements are true or false. Correct the false statements. R
1. Martin helps young criminals.
2. Many teenagers commit a crime because of problems at school.
3. Martin explains some harmful effects of drugs.
4. Many young people take drugs for a long time.
5. Some young people think drugs could help them with their problems.
6. Drug-taking may be a response to loneliness or result from low self-esteem.

I **c** Role play:

- Student A is a social worker at college.
- Student B is a student with a drug problem.
- Student C is student B's class teacher.
- Student D is student B's father/mother.

Prepare and act out a drug counselling session in which everyone depicts student B's problems and then try to come to a solution to the problem.

P **d** Now write a report on your session. → SF 7, 8, 13

P **e** Design an anti-drugs poster and present your results to your class.

Immigration in the United States

Two hundred years ago the U.S. had a population of about 4 million people.
Today the population has risen to about 300 million.

Immigrants – Where do they come from?

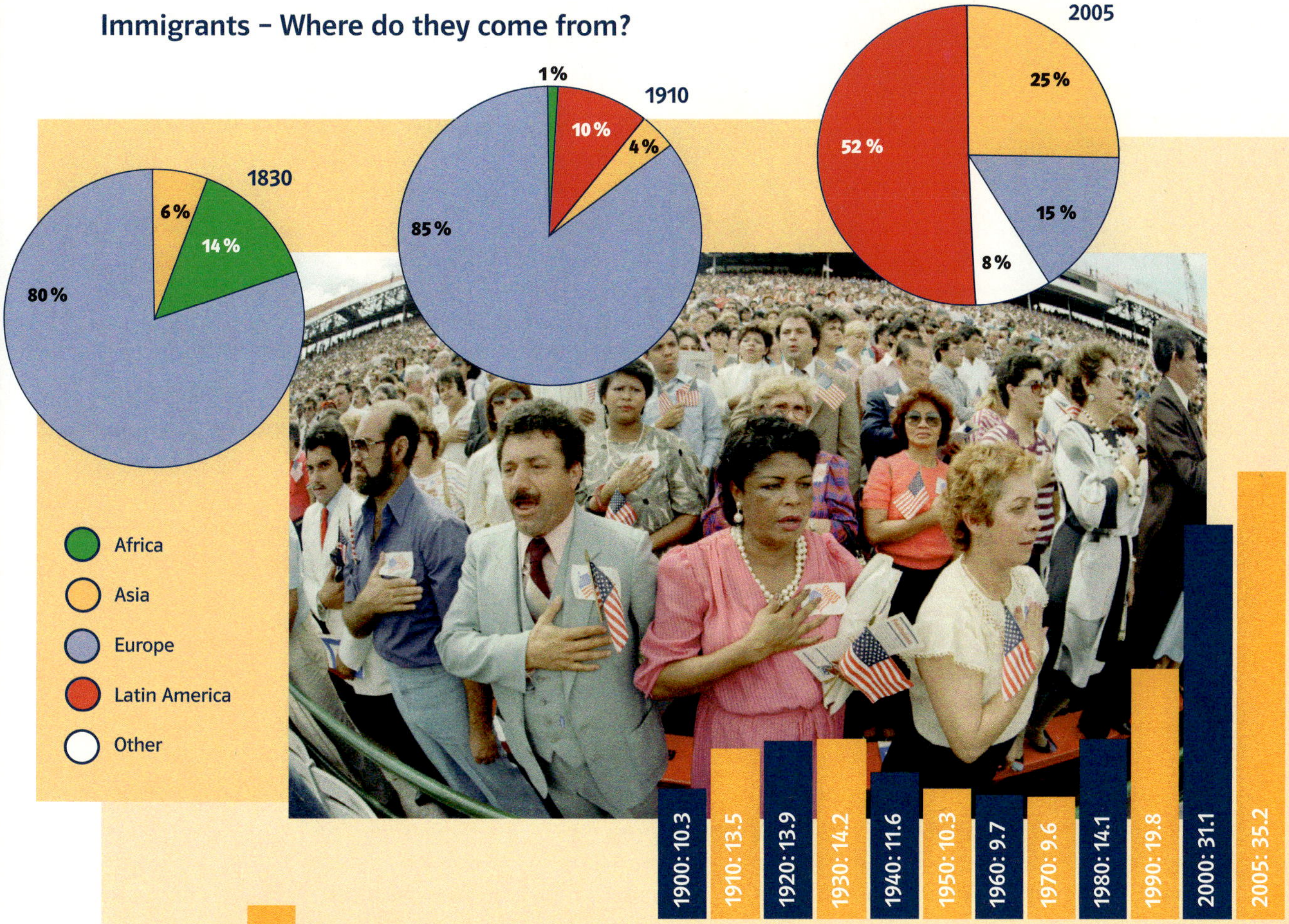

Total number of immigrants (in millions)

The development of U.S. immigrant population 1900 – 2005

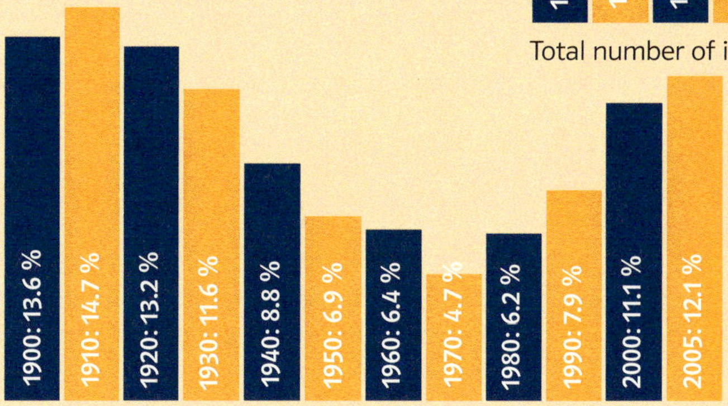

Immigrants as percentage of U.S. population 1900 – 2005

a Describe the development of immigration in the United States. **→ SF 21**

b Where have the different groups of immigrants come from over the years?

c What reasons have they had for coming? What do you think?

Mexico's dilemma – Jobs could solve some problems

(A) Of all the ways to limit unauthorized human traffic across the USA's southern border, there's one remedy that few people are even discussing: creating more jobs in Mexico. The Mexican economy in 2006 is expected to have its third-consecutive year of better than 3% growth, but the jobs situation isn't as good. Each year, about 1 million Mexicans enter the labor pool. But the economy creates only a small number of the new jobs needed for them. It doesn't require a degree in economics to understand that if there were more well-paid jobs in Mexico, fewer Mexicans would enter the USA in search of work. Today, more than half of the roughly 500,000 unauthorized immigrants entering the USA each year are from Mexico, according to the Pew Hispanic Center.

(B) So why doesn't the Mexican economy — more than a decade after a free trade deal with the United States — produce more jobs? "Although Mexico has done all the right things at the macroeconomic level … and has integrated with the U.S., it has not done the last fundamental reforms," says Claudio Loser, a former International Monetary Fund (IMF) official. Despite dropping trade barriers and liberalizing its financial markets, Mexico remains economically stuck. The consensus view is that to produce an adequate number of jobs, the USA's southern neighbor requires far-reaching reforms of labor markets, investment policy and the tax system — almost every critical component of a world-class economy.

(C) However, Mexico's mission isn't impossible. Other countries in Latin America and elsewhere have pulled themselves from poverty to prosperity. From 1980 to 2004, both South Korea and Ireland roughly tripled per-capita incomes while Mexico saw little more than a 10% jump. Today, just 12.8 million of the country's 43 million working-age population have jobs in the formal sector, according to the IMF. (Millions more work in irregular jobs with low or uncertain wages and no benefits.)

(D) Unless job-creating reforms are carried out, the enormous gap in living standards on opposite sides of the border will continue to draw Mexican labor into the USA. Today, the ratio between American and Mexican incomes is roughly 6-to-1. To limit illegal immigration, Mexico doesn't have to become as rich as the USA. It just needs to close the gap and give its citizens hope of further progress. After all, Connecticut's per-capita income is almost twice that of Mississippi's. Yet plenty of people live in the Magnolia State.

(E) Mexico's dilemma, however, is growing more urgent because of the rise of China. Since joining the World Trade Organization (WTO) in 2001, the Asian power has emerged as the low-cost supplier for industries from furniture to toys. Mexico, after years of operating as the low bidder, suddenly finds itself priced out of many markets. The two developing nations compete head-on in about one-half of the goods — equal to three-quarters of their total export value — that each ships to the USA, the IMF says. Manufacturing costs are almost twice as high in Mexican clothing plants as in their Chinese rivals. In the three years ended in 2003, Mexican textile job losses totaled 187,000. "Mexico is not well-positioned in the major dynamic markets," concludes the IMF. During the

NAFTA (North American Free Trade Agreement) debate, Mexico's then-president Carlos Salinas said the trade deal would allow Mexico to export "tomatoes, not tomato pickers". Yet, more than a decade later, the tomato pickers keep coming.

(F) By the end of this decade, demographics should cool the immigration controversy. Mexico had the world's highest birth rate in the 1960s, when the typical woman had seven children. Now, the average is down to 2.6 children. So the annual number of people turning 15 years old will drop from a peak of about 1 million in 2003 to about 600,000 by 2010, according to Philip Martin, a professor at the University of California, Davis. "Demography is real. There will be less migration pressure from a rapidly growing Mexican labor force in the future," Martin says. But unless Mexico finds the political will for comprehensive economic reforms, cross-border flows won't stop entirely. What would be needed to really make a dent in illegal immigration? "The economy would have to grow at double the rate it is today," Loser says. "It may take a generation."

730 words

(USA TODAY, May 2006)

P **a** Describe the photo on page 82. → SF 19

b With regard to the heading of the text above – what might the photo illustrate?

c Scanning the text: Read the text quickly and find the following items: → SF 2

1. Five countries
2. Three international organisations
3. The following four numbers and the information that they give:
   a) five hundred thousand
   b) forty-three million
   c) six to one
   d) seven

R **d** Multiple choice: Find the most suitable option.

1. Mexico's economy has been growing
a) by more than 3% in the last 3 years.
b) by 3% in the last 3 years.
c) by more than 3% since 2001.
d) by almost 3% in the last 3 years.

2. Economists suggest that the Mexican economy isn't creating enough jobs because
a) trade barriers were removed.
b) there were no reforms of the financial markets.
c) there were no reforms of the labor market.
d) too many skilled workers leave the country.

3. The IMF says that
a) Mexico and China compete in about three-quarters of their goods.
b) Mexico and China compete in about 50% of their goods.
c) Mexico ships 75% of its export to the USA.
d) Mexico and China ship 75% of their export to the USA.

4. There will be less migration pressure in the future
a) as Mexico's economic growth will be stronger.
b) as fewer young people will enter the Mexican labor market.
c) as more jobs will be created in Mexico.
d) as there will be more financial aid for Mexico's poor.

M **e** Übertragen Sie die Abschnitte A und B ins Deutsche. Übersetzen Sie nicht Wort für Wort und schreiben Sie ganze Sätze. → SF 10

P **f** Find further information about the history and the tasks of the three international organisations mentioned. Present it to your class. → SF 13

1 California and the Chinese

The Californian population is a mixture of immigrants from many different nations **which includes** lots of Chinese. Most foreigners know the part of San Francisco **which is called** Chinatown. Today Chinatown has not only a lot of fine restaurants **which have been built** for the tourists, but it is also a place where some Chinese people do not have enough to eat. In some areas there are families with four or five children **who live** in a single room. And so you find many families **who use** the same kitchen and bathroom. Today Chinatown is not only a tourist attraction but also a place with a lot of poverty. Meanwhile many young Chinese have left Chinatown for other places **which offer** better jobs. But most of the new immigrants **who come** from China cannot move because they do not know the language **which is spoken** outside the Chinese community. The older Chinese people **who have been** there for decades do not want to move to another place. They hold on to their old ways, and they do not often ask for help from outside the community.

Replace the words in bold type by a present participle (ing-form) or past participle (third form of the verb).

Example:
The Californian population is a mixture of immigrants from many different nations **including** lots of Chinese. Most foreigners know the part of San Francisco **called** Chinatown. Today Chinatown …

2 Native Americans in California

The American Indian was the first true Californian. The style of Indian art, music and clothing is part of the whole West, including California. But like other minorities, the Indians have had trouble in Californian society. In their case the problem is land. **As they believe** that they have been treated unfairly by the American politicians, many of them have become rather bitter. Others have started to make their voices heard in various protest actions. One group includes the Pitt River Indians. **Before they started** their protest action, they talked to several reporters.

Now put the following statements about the Pitt River Indians into the same form as in the examples below.

1. When they realized how much gold there was on their land, the whites decided to take it away from the Indians.
2. As they knew the Indians had been treated badly, later governments gave them money for their land.
3. Because the Indians want their land back and not the money, they have decided to start a protest action.
4. After they have made their protest here, they are going to fly to Washington.
5. As they know how difficult it is to get a meeting with the President, they are prepared to wait as long as necessary.
6. Before they leave Washington the Indians will give a press conference.

Example:
Believing that they have been treated unfairly … **Before starting** their protest action …

3 A Mexican immigrant

A Mexican immigrant talks about his life in the U.S. during a break in his work.

■ Infinitive or ing-form? Fill in the correct form of the verb.

Example:

1. I'll never forget leaving my small Mexican hometown.
2. My mother told me: "Don't forget to write to us …"

I'll never forget **1. (leave)** my small Mexican hometown. My mother told me: "Don't forget **2. (write)** to us when you're there." I remember **3. (cross)** the border by night. I was afraid of **4. (be arrested)**, but I wanted **5. (change)** my life completely, so I decided **6. (take)** the chance. In the first town I stopped **7. (get)** some food. Inside the restaurant there was a sign **8. (say)** that they needed help in the kitchen. I remember **9. (ask)** the boss what he was prepared **10. (pay)**. It wasn't much, but **11. (need)** the money urgently I accepted it.
They made me **12. (do)** the dirty jobs. First I enjoyed **13. (work)** there because the other workers were friendly. But after **14. (stay)** there for six months I moved here to San Diego. I hope **15. (find)** my own apartment soon. I would like **16. (stay)** here forever. That's why I must remember **17. (phone)** the Housing Officer tomorrow. I used **18. (be)** a so-called illegal immigrant, but a couple of years ago I got a green card. Now I'm able **19. (live)** and work wherever I like.
– Sorry, my break is over. I must stop **20. (talk)** now. I have **21. (get)** back to work again.

4 A German immigrant

Ursula Zimmermann from Germany emigrated to America twenty years ago, but she has always kept in contact with her friends in Germany.

■ Translate the following part of one of her e-mails. → SF 10

Vor einigen Monaten sah ich öfters drei junge Leute im Supermarkt. Ich erinnere mich, dass ich Herrn Cheng, den Leiter des Supermarktes, gefragt hatte, wer sie waren. Er erzählte mir, dass sie deutsche Studenten waren, die eine Reise von Kalifornien nach New York planten. Ich erinnere mich, dass ich sie eines Abends traf. Sie reparierten gerade ihr altes Auto. Ich hielt an, um mit ihnen zu reden. Es machte mir richtig Spaß, wieder Deutsch zu sprechen. Die jungen Leute freuten sich darauf, die USA zu durchqueren und die vielen Sehenswürdigkeiten zu sehen. Ich bot ihnen an, bei der Planung der Reise zu helfen.

Immigration and welfare

Together with corn and cars, immigration has been one of the main factors of American economic growth. The traditional theory is simple: new workers increase the supply of goods and services with their labor and increase the demand for other goods and services by spending their wages. A circle of growth begins as a growing number of workers create a richer society for each other. Two hundred years of U.S. history seem to confirm this theory. But there is a feeling today that immigration is not so good for the economy any more.

The polls show that many U.S. citizens are worried that more immigrants will take jobs away from native-born workers. This can be true in times of high unemployment. In California, for example, a recent survey shows that the population explosion between 1990 and 2000 was the result of a massive inflow of immigrants. With a jobless rate close to seven per cent (2003) many native-born Americans are actually leaving to find work in other states. So we must see both sides: "The short-term costs of immigration are much higher today," says Michael Boston, one of the leading American economists, "but in the long run, immigrants are still great news for our economy. Normally, new jobs are created through the immigrants' own work. The immigrants' spending creates a demand for houses, food, etc. and their employers invest their growing profits in new machinery and jobs. That's how America got rich."

However, in the last few decades things have changed. In the early days of immigration public education and some public health programs were the only services for those coming to New York or other Northeastern cities. One third of the new immigrants soon found out that America did not offer the opportunities they were looking for and so they moved back home. Nowadays – even after the welfare reforms in 1996 which restricted help for immigrants – a lot of different welfare programs, from food stamps to unemployment benefit, help those who do not succeed in finding work and even attract immigrants who would otherwise stay in their home countries. Furthermore, the level of skills of 90 per cent of the new immigrants is relatively low. And so are their earnings. As a result, state governments spend between $11-22 billion to provide welfare to immigrants, and welfare costs are steadily climbing. Today welfare use by immigrants is 43 percent higher than by native-born Americans and immigrants are still entering the United States in high numbers. In many states immigration is said to be one of the main reasons for their financial problems and many experts believe that unless American immigration policies are changed there will be no end to the problem in the near future.

457 words

R **a** Read the text quickly – as if you were reading a newspaper article – and say in short what the text is about. Then find out which of the following topics are mentioned in the text. → SF 2

unemployment • transport • welfare costs • California • Mexico • crime • skills • Texas • welfare use • environment

R **b** After you have read the text carefully again, answer the following questions. Use your own words as far as possible.
1. How did the American economy benefit from immigration in the past?
2. Why do many Americans see immigration as a threat?
3. In which way are things different today than a few decades ago?
4. Which effects does immigration have on the budget of the states?

M **c** Vocabulary
Explain the following words in English.
1. welfare **(headline)**
2. wages **(line 10)**
3. polls **(line 19)**
4. native-born **(line 21)**

d Translate lines 44 to 70. → SF 10

I **e** Role play: Discussion → SF 14
"Immigration helps the economy." Do you agree? Be prepared to discuss this by taking over one of the following roles. Follow the steps in the box below.
1. You are an immigrant from India. You are 25 years old and a qualified computer programmer and live in America now.
2. You are 45 years old, born in the USA and now an unemployed bricklayer.
3. You are an American manager of an IT company.
4. You are an American social worker in Chicago.

P **f** Presentation: Immigration in the United States → SF 13
Prepare a presentation of this topic, using the information you have learnt from unit 10. If possible, use other sources of information such as the Internet. Then give a presentation in class.

P **g** Write a comment on the following statement, using your own knowledge and ideas.
"Immigration brings a lot of benefits for our economy." → SF 5

1 The new majority

The Californian lifestyle has grown out of many different cultures. People have been coming here from other American states and from foreign countries for a long time now. They have come to find better jobs, a better house, a better climate – in short a better way of life.

R **a** Listen and put the photos into the correct order.

R **b** Find out where these people came from.

R **c** Find out where they live now in California.

R **d** Listen again and find out the following information about the people:

	Marital Status	Language	Job in U.S./home country	Reasons for emigrating	Problems
Santiago					
Lev Pischnik					
Mom Chay Suth					

2 Minority groups

In many countries minority groups often face problems. Therefore some people are against immigration. On the other hand, immigration brings a lot of benefits to a country.

P **a** Make a list of all the benefits you can think of in the following areas.

1. economy
2. employment
3. welfare
4. education
5. culture
6. language

b A local newspaper has just published a letter from someone wanting to stop immigration. Write a letter to the editor of the newspaper, giving your arguments for immigration. → SF 5, 7, 8 P

Special topics

Intercultural studies

World of business

The American way of life

1 From America with love – A letter to a girlfriend

a Describe the photos above.
What are the differences compared to Germany?

R **b** Read the text on page 91–92 and make a list of the things that Esther found different in America.

Language	Food and drinks	Clothes	Behaviour	Attitude	School

P **c** Describe the differences in complete sentences.

Dear Julia,

How are you? I'm fine! I've been with my American host family in Duncan, South Carolina, for almost four weeks now and I'm enjoying my stay a lot. We get along very well, although there are more differences than I had expected. It started on the evening of my arrival: After my family had picked me up from the airport, we had supper at home. I was surprised nobody said "good appetite" or something like that before the meal. We ate a whole chicken with sauce. There were no potatoes or rice, no vegetables, not even bread. I liked the chicken, but I was surprised that the meal only consisted of meat! Now I've learned that for an American family, that is not extraordinary. We eat a warm meal like meat loaf with mashed potatoes or pepper steak with rice every night. The whole family usually drinks a can of coke or another soft drink with the meal. There is always a big box of cans in the fridge instead of bottles like they are used in Germany. Well, after my first supper with the family I experienced the next difference. I wanted to know where the toilet was and they didn't seem to understand until someone said, "Oh, you mean the bathroom?" In the United States, you NEVER say "toilet" although, in fact, that's what you mean! That's one of those essential things they never teach you in English class… It's also important to know that it's considered impolite to blow your nose at table or in the classroom. You either go to the "bathroom" or into the hall which is quite unpractical when you have a cold!

I also noticed that Americans never say "trousers", but "pants". But since almost everybody wears jeans that is the word you hear most of the time! What also surprised me is that everybody is so polite! When you go to the supermarket, the cashier doesn't just say "Hi", but says "Hi, how are you?". I'm always tempted to say, "Thanks, I'm fine", but I've learned that it's better not to answer, but to ask the same question in response.

I also have to get used to going to church twice a week. We go there every Sunday morning and almost every Wednesday night. My host family belongs to a baptist church, which is a Christian protestant church. On Sundays, service begins at 10.30 and an hour earlier, the church offers a class called "Sunday School" where you discuss religious topics and learn how to be a good Christian. The Baptist service is not very different from a Protestant service in Germany, but contrary to a German church, there are always a lot of people attending the service, including many children and teenagers.

I have the impression that the church is not only a place of worship but also a place for meeting friends.

My host family is quite religious as are many American families. They usually "say grace" before the meals, they do not smoke or drink alcohol and do not use swear words. Furthermore, young couples usually aren't allowed to spend the night together or even close the door of their room when they are together.

This rather conservative attitude also becomes apparent at school. Girls are not allowed to wear t-shirts with tiny straps (they have to be at least three fingers wide), short skirts or clothes that expose the stomach. You are not allowed to wear bandanas, hats or sunglasses. If a couple hold hands, touch or even kiss at school they can be punished for PDA (public demonstration of affection). And these are just some examples! In general, I have noticed that students are more disciplined in the classroom and that there is a quieter atmosphere. On the other hand, you are not forced to participate because only written tests count for your grade.

As a consequence, there are students who go to sleep during a class! I have to admit that I am quite disappointed that we don't have many discussions in class. You usually just listen to the teacher or do written tasks. Even in language classes, you hardly speak! But how can you learn a language without ever speaking it?
By the way, my school's called James F. Byrnes High School. I'm in twelth grade here, so I'm called a senior. I might even be able to take the final exams at the end of the school year because in the US, you graduate after twelth grade. My subjects during this term are Spanish, US History, Biology and Orchestra. It was hard to pick just four because there are so many different subjects offered! I have those same four every day in the same order. That's a little boring but you never forget your schedule! Next term I can choose Driver's Education as a subject. Maybe I can get my license. That sounds cool!
I was astounded when I was at school for the first time and all the students got up to say the "Pledge of Allegiance". In the morning before the first class starts, you get up, put your right hand on your heart, turn towards the flag (there is one in every classroom) and pledge allegiance to the American flag and nation. At first, I didn't know what to do because I had never seen such a thing in Germany and I hadn't even heard of it before!

Well, there are many more things I could tell you about, but I have to do the dishes now… it's my turn today!
I'm looking forward to hearing from you!

Love,
Esther

940 words

2 British English – American English

a Read the words in the two boxes and find out which box contains American English and which one British English? How do you know?

b Match the words from the left with suitable words from the right.

1. freeway	A. holidays
2. soccer	B. bank note
3. baggage	C. chips
4. closet	D. flat
5. gas	E. toilet
6. trunk	F. motorway
7. cookies	G. football
8. rest room	H. car
9. sidewalk	I. wardrobe
10. auto(mobile)	J. petrol
11. fall	K. boot
12. vacation	L. biscuits
13. bill	M. autumn
14. French fries	N. pavement
15. apartment	O. luggage

c What are these holidays called and when are they?
Match the photos with the holidays. Then match the holidays with the dates.

1. Labor Day	a) 4 July
2. Thanksgiving	b) 1st Monday in September
3. Independence Day	c) 4th Thursday in November

A

C

d How are the following words written in American English?

colour • centre • analyse • catalogue • programme • licence • travelling • tyre • cheque • dialogue • fulfil • favourite • theatre

P **e** Look at the home page of James F. Byrnes High School on the internet. Present the school to your class. Talk about similarities and differences compared to the German school system. → SF 13

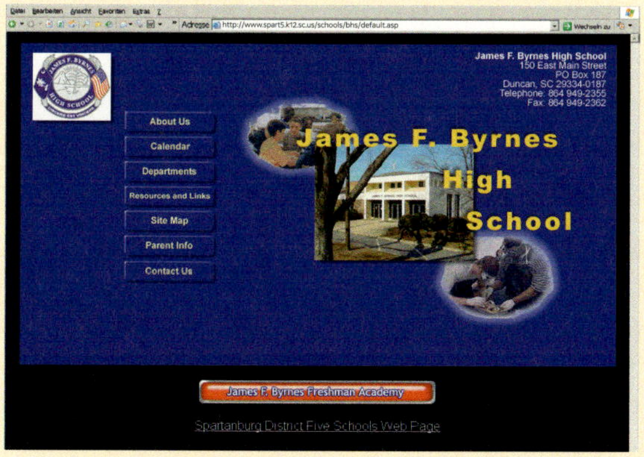

P **f** **What about you?** Have you ever been abroad on your own or together with your class? Where? What differences did you experience? How did you react? Prepare a short talk to your class. → SF 13

Cross cultural communication

1 Meeting someone for the first time

When you meet someone for the first time, you would shake hands. But after this first meeting people in English-speaking countries do not normally shake hands again.
Moreover, in English we often use first names. This does NOT mean that when you use a person's first name that it is the same as when you use "Du" in German. In English you can use a person's first name and still be quite formal.

■ **When Martin Hope of Singapore Trading comes to Germany on a business trip, you meet him for the first time. What would you say to him in each of the following situations? One phrase in each group is not possible. It may be incorrect, impolite or inappropriate. Say why.**

1. Meeting and greeting
   a) Hello, Mr Hope, nice to meet you.
   b) Good afternoon. How do you do? Did you have a good trip?
   c) Hi, Martin, how are you? Did you have a good trip?
   d) Mr Hope! I'm Greta Fabian. I'm pleased to meet you.
2. Introducing others
   a) This is Hanna Freese, our departmental assistant.
   b) May I introduce you to Hannes Strauss? He works in the Accounts Department.
   c) I'd like you to meet Tom Wolff – he's the marketing manager.
   d) That's Peter Dotter. Do you really want to meet him?
3. Looking after a visitor's needs
   a) Now, would you like to go to the toilet?
   b) Have you got all you need?
   c) Would you like to freshen up?
   d) Is everything OK with the hotel?
4. Checking through the schedule
   a) I suggest that we deal with … tomorrow.
   b) You won't mind finding your own lunch, will you?
   c) We could meet Mr Strauss for lunch and for an initial discussion.
   d) Directly after lunch we could have a look at the plans for …

2 Using polite expressions

When talking to English-speaking people you will find that they often use a lot of polite expressions. It is important in the English-speaking world to use such expressions. Otherwise you may sound unfriendly or even impolite.

Helpful phrases.

Please	I'd like …
Thank you. – You're welcome!	Here you are!
I'm sorry	Would you mind if …?
Excuse me, could you (possibly) …?	Would you be so kind as to …?
Pardon?	

Role Play:

You are in a restaurant in Ireland. One student plays the guest, the other the waiter/waitress.
The waiter opens the dialogue by asking what he can do for the guest. The guest wants the menu. After reading the menu, the guest gives his order to the waiter/waitress. He /she takes the order. Later the guest wants the bill and pays. Guest and waiter/ waitress say goodbye.

3 Small talk → SF 15

We use small talk in a variety of situations, but especially when we do not know the person we are talking to very well. The important thing about small talk is that we shouldn´t become too emotional or intense. We must make sure that we are just being polite.

Helpful phrases.

Talking about where you come from
What part of … do you come from?
I'm from … That's near …
Do you know …?
Have you been to … before?

Talking about journeys
Did you have a nice journey/flight/drive?
Did you have any difficulties finding us/getting here?
How did you get here?

Talking about accommodation
Where are you staying in …?
Are you happy with your accommodation/the hotel?
How is the hotel where you are staying?

Talking about your family
Are you married, by the way?
Do you have any children?
How old are they?

Talking about leisure time
What do you do in your free-time?
Are you interested in …?
Do you like …?
Do you do any sports?

Talking about holidays
Where do you like to go on your holidays?
Have you ever been to …?

There are of course many other topics of conversation. What you talk about depends very much on how well you get on with the person you are talking to. However, as a rule you should avoid talking about politics, religion, personal problems or illnesses.

Role Play:

You are at a trade fair in Birmingham and you have been invited to a promotion party. During the party you meet people from all over the world and talk to them.
Act out this scene. Walk about in your classroom and do some small talk (e.g. about the weather, your journey to Birmingham, your family, sport and hobbies, your holidays) with one or two of the other students. After two minutes turn to a new partner/new partners.

4 How should you behave in business?

R ■ Complete the following sentences using one of the three options.

1. When meeting people for the first time in Britain you should
   a) talk loudly.
   b) stand a short distance away from the person you are talking to.
   c) kiss or hug the person.
2. When on business in South America you should
   a) be on time.
   b) keep in eye contact.
   c) stand well away from your business partner.
3. When meeting British business colleagues you should
   a) shake hands only with the women.
   b) shake hands with them.
   c) shake hands only on leaving.
4. When you are at a meeting with Japanese partners you should
   a) give a present to your business partners.
   b) stare at your business partners.
   c) touch your partner.

5 Cross cultural negotiation training

R a Read the text and check if your answers in 4 are correct.

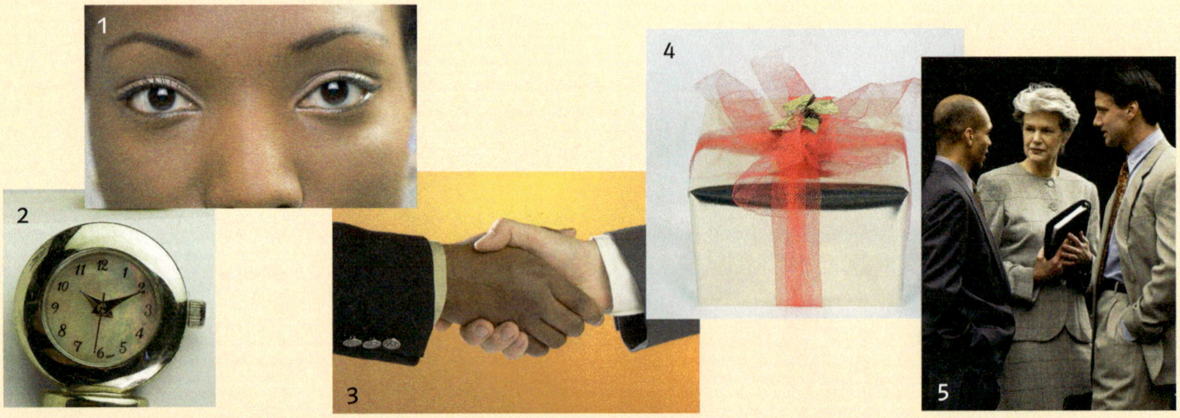

Cross cultural negotiation is one of many specialized areas within the wider field of cross cultural communications. By taking cross cultural negotiation training, negotiators and sales personnel give themselves an advantage over competitors.

Let us look at a brief example of how cross cultural negotiation training can benefit the international business person:

There are two negotiators dealing with the same potential client in the Middle East. Both have identical proposals and packages. One ignores the importance of cross cultural negotiation training believing the proposal will speak for itself. The other undertakes some cross cultural training. He/she learns about the culture, values, beliefs, etiquette and approaches to business, meetings and negotiations. Nine times out of ten the latter will succeed over the rival. Cross cultural negotiations is about more than just how foreigners close deals. It involves looking at all factors that can influence the proceedings. By way of highlighting this, a few brief examples of topics covered in cross cultural negotiation training shall be offered.

Eye Contact:
In the US, UK and much of northern Europe, strong, direct eye contact conveys confidence and sincerity. In South America it is a sign of trustworthiness. However, in some cultures such as the Japanese, prolonged eye contact is considered rude and is generally avoided.

Personal Space & Touch:
In Europe and North America, business people will usually leave a certain amount of distance between themselves when interacting. Touching only takes place between friends. In South America or the Middle East, business people are tactile and like to get up close. In Japan or China, it is not uncommon for people to leave a gap of four feet when conversing. Touching only takes place between close friends and family members.

Time:
Western societies are very "clock conscious". Time is money and punctuality is crucial. This is also the case in countries such as Japan or China where being late would be taken as an insult. However, in South America, southern Europe and the Middle East, being on time for a meeting does not carry the same sense of urgency.

Meeting & Greeting:
Most international business people meet with a handshake. In some countries this is not appropriate between genders. Some may view a weak handshake as sign of weakness whereas others would perceive a firm handshake as aggressive. How should people be addressed? Is it by first name, surname or title? Is small talk part of the proceedings or not?

Gift-Giving:
In Japan and China gift-giving is an integral part of business protocol, however, in the US or UK, it has negative connotations. Where gifts are exchanged should one give lavish gifts? Are they always reciprocated? Should they be wrapped? Are there numbers or colours that should be avoided? All the above in one way or another will impact cross cultural negotiation and can only be learnt through cross cultural training. Doing or saying the wrong thing at the wrong time, poor communication and cross cultural misunderstandings can all have harmful consequences.

Cross cultural negotiation training builds its foundations upon understanding etiquettes and approaches to business abroad before focusing on cross cultural differences in negotiation styles and techniques. **524 words**

(Cross cultural Negotiation, The Sideroad, Blue Boulder Internet Publishing, Neil Payne, 2005)

p **b** Describe what you can see in the photos. → SF 19

c Put the photos in the order they appear in the text.

d Complete the following sentences using words from the text.
Very often in Europe and America when you look your business partner in the eye, they have … **(1)** … in you. In Japan, on the other hand, if you look a person in the eye for a … **(2)** … time, they … **(3)** … this as impolite. So you should generally … **(4)** … doing that in Japan. Furthermore, Japanese people do not like to stand too … **(5)** … to each other when talking. And they would only … **(6)** … people they know very well. And one more thing: don't come late to a meeting in Japan ; that would be an … **(7)** … to your host there.
And in Japan don't forget to take a … **(8)** … with you for your host.

M **e** Mediation → SF 10
Begründen Sie, warum ein Geschäftsmann oder eine Geschäftsfrau mehr Erfolg erzielen kann, wenn er/sie „cross cultural negotiation training" betreibt.

Job application

1

MHS PLASTICS (REF.: ET 102)

We are seeking enthusiastic individuals from the Continent to work in our Manchester-based export team.
Applicants must be educated to A Level standard or equivalent and have excellent communication skills. Native speaker competence in Italian, French or German and good English language skills are required.
An initial one year package is on offer including comprehensive training, good rates of pay, challenging and interesting work leading to a permanent position.

Send application enclosing CV and certificates in English to:
Robert Mills, Recruitment officer, MHS Plastics Ltd,
67 High Street, Manchester, M1 2HF

2

Travel Reservations Agent

Are you interested in travel reservations?
This travel reservations company based in London are looking for motivated reservations agents.

Reservations Agent responsibilities:
- Making reservations of flights (chartered or scheduled), car hire, insurance and other add-ons
- Giving excellent customer service
- Pro-actively selling all travel extras

Skills required:
- Fluency in German or French, good command of English
- Good worldwide geographical knowledge
- Previous experience in travel reservations preferred (training available)

If you would like to apply, please send your CV with a covering letter to scott.craventravel@candm.co.uk or call Scott on 02073972270 quoting ref SP927

1 Job advertisements

R **a** Compare the two advertisements and answer the following questions.
1. Which advertisement appeals to you most. Why?
2. What kind of job do the two companies offer?
3. Would you apply for one of the jobs offered? Say why or why not.
4. How can you apply for one of the jobs?

I **b** Discuss what German qualifications would be needed for the two jobs advertised.

c Describe your (future) qualifications (eg school leaving certificate, apprenticeship, traineeship, language certificates) and state briefly what kind of job you would like to apply for.

2 Letter of application

Info: What makes a good letter of application?
You should always include a letter of application when you apply for a job. It is a selling tool that outlines your skills and experience so that your prospective employer can see what you have to contribute to the company or organisation. The following tips can help you to write a good letter of application and to finally get an invitation to an interview:

1. No typing or spelling mistakes.
2. Address your letter to the person who is making the hiring decision.
3. Don't simply copy the form and style of a standard letter. Instead you should individualize your letter by showing how your individual qualities can contribute to the company.
4. Show that you know something about the company and the industry.

Intercultural communication
In English letters it is frowned upon to use an impersonal salutation like "Dear Sir or Madam". Instead you should always try to find out the name and correct title of the person you are writing to. Use "Mr" for all men and "Miss" for unmarried females, "Mrs" for married or divorced females and "Ms" for females where the marital status is unknown. Nowadays in the world of work the neutral form of address "Ms" is often used for all women.

a Match the German occupations with their English paraphrases.

1. Automobilkaufmann	a. Management Assistant in Informatics
2. Bankkaufmann	b. Management Assistant in Event Organisation
3. Bürokaufmann	c. Insurance Business Management Assistant
4. Fachkraft für Lagerwirtschaft	d. Management Assistant in Hotel and Hospitality
5. Hotelkaufmann	e. Management Assistant in Warehousing and Logistics
6. Informatikkaufmann	f. Automobile Sales Management Assistant
7. Kaufmann im Einzelhandel	g. Management Assistant in Sports and Fitness
8. Reiseverkehrskaufmann	h. Travel Management Assistant
9. Sport- und Fitnesskaufmann	i. Management Assistant in Retail Business
10. Veranstaltungskaufmann	j. Bank Business Management Assistant
11. Versicherungskaufmann	k. Office Management Assistant

M **b** What would you say in these situations?
1. Ich bewerbe mich um einen Ausbildungsplatz als Versicherungskauffrau.
2. Ich möchte gerne eine Ausbildung als Veranstaltungskaufmann machen.
3. Ich habe bereits ein Praktikum in einer Bank gemacht.
4. Ich mache eine Ausbildung als Einzelhandelskauffrau.
5. Ich möchte gerne Sport- und Fitnesskaufmann werden.

P **c** Write a letter of application for one of the jobs above. → SF 11
Use your own personal details.

3 Curriculum Vitae (CV)

A CV is an important part of any application. German job titles, types of schools and diplomas often have no direct equivalent in English. Therefore, you should use paraphrases so that a potential employer has a clear and realistic impression of your education and experience. When applying for a job in English be aware of differences.

Intercultural communication
In an English CV you should not give any details of your parents' profession, your religion, or of the number of your brothers and sisters. You should also not include a photo when applying in the USA or Britain.

Curriculum Vitae

Personal information	
Name and address	Simone Wegener Schwarzwaldstr. 10 68163 Mannheim, Germany Tel. + 49 (0)621 832752 Email: simone.wegener@t-online.de
Nationality	German
Date of birth	20 April, 1988
Place of birth	Heidelberg, Germany
Marital status	single
Desired employment	Travel agent
Education	
1994 – 1998	Primary School, Heidelberg
1998 – 2004	Humboldt Realschule (Higher Secondary School), Mannheim
2004 – 2006	Max Hachenburg Berufskolleg (= Higher Secondary Vocational School), Mannheim
2006	Fachhochschulreife (examination enabling student to enrol at a polytechnic university) – Overall grade: 2.3 (= good = B)
Work experience and other activities	
September 2004 – June 2005	Junior volleyball coach
August 2005	Internship at the "Flypoint" travel agency in Mannheim
July 2006 – October 2006	Temporary employment at the "Flypoint" travel agency in Mannheim
Languages	German: native command English: good command (spoken and written)
Computer skills	Windows, Word, Excel, PowerPoint
Personal interests	Volleyball, painting

a Match the beginnings of the the following tips for writing a CV with the correct endings.

1. Structure	a. or paraphrase school types, diplomas etc.
2. No reference to	b. skills that might be valuable for your potential employer.
3. Include	c. desired employment.
4. Don't sign your	d. be asked about them.
5. Think carefully about your interests. You might	e. name at the end.
6. Include language and other	f. your race, religion or family.
7. Give German equivalents	g. your CV clearly
8. Do not include	h. a photo.

P **b** Study the model CV above and write your own CV. Use a dictionary if necessary.

4 Job interview

Preparing for an interview

Preparation is a key part of any interview process. However, many do not commit enough time. Remember, you only have one chance to make a good first impression! Five steps to follow in order to be properly prepared include:

1. Know your employer – Find out as much as possible about the company: its products, its structure and if possible recent achievements. This will not only increase your self-confidence, but show your real interest in the organisation.
2. Know yourself – Evaluate yourself in terms of strengths, and how your future employer could use them. If you need to state a weakness, do not elaborate on it and try to turn it into a potential strength. Think about your interests carefully. If you mention an interest in your CV, you should expect to be asked about it.
3. Communicate effectively – If you can't communicate your knowledge clearly, your interview will not be impressive. The best way to improve your communication skills and self-confidence is to practice role-playing beforehand.
4. Be careful about the so-called "soft skills" – If you are looking for a professional position, you must look like a professional. Therefore, a neat, clean appearance, polished shoes and a neat hairdo are essential. Women should wear a simply tailored suit or dress and men should go for a suit with a pressed shirt and tie. Remember: The first impression is a lasting one so don't risk wearing a pair of jeans and a t-shirt and avoid flashy jewellery.
5. Finally, be on time – It's always better to be a few minutes early than one minute late for your interview. So make sure you know how to get to the interview and how long the journey takes.

R **a** **Answer the following questions.**

1. Why is it important to know something about the company?
2. Why should you think about your strengths and weaknesses?
3. What interests or hobbies would you mention?
4. What is meant by the term "soft skills"?
5. Can you think of anything else that is important in an interview?

R **b** **Listen to the interview and write down the job details mentioned.**

R **c** **Listen to the interview again and translate the following sentences into English.**

1. Sie haben ein Praktikum bei der Kundenserviceabteilung eines Online-Händlers gemacht.
2. Unsere Deutsch sprechenden Kunden brauchen jemanden, an den sie sich wenden können, falls es Probleme gibt.
3. Wir werden Ihnen davor eine Schulung geben.
4. Wann kann ich erwarten, von Ihnen zu hören?
5. Ich habe mich gefreut, Sie kennen zu lernen. Sie werden in Kürze von mir hören.

I **d** **After you have sent off a letter of application and your CV for one of the jobs advertised on page 98, you are invited for a job interview. Work with a partner and act out the following interview as a role play in English. You may take notes first.**

Partner A (Interviewer):
- Beginnen Sie mit allgemeinen Fragen, z.B. nach Flug, Heimatstadt des Bewerbers.
- Fassen Sie die Anforderungen für die Stelle kurz zusammen.
- Fragen Sie nach Qualifikationen, z.B. Software-Kenntnisse.
- Erkundigen Sie sich, wieso der Bewerber in GB arbeiten will.
- Gehen Sie auf die im CV genannten Interessen kurz ein.
- Versprechen Sie in einer Woche Bescheid zu geben und verabschieden Sie den Bewerber freundlich.

Partner B (Interviewee):
- Beantworten Sie die Fragen des Interviewers ganz nach Ihrem Wunsch bzw. auf Basis Ihrer Bewerbung (Letter of application and CV).

e **Find the equivalents for the following German school types in the box below.**

1. Hauptschule
2. Realschule
3. Gymnasium
4. Berufsschule
5. Berufsfachschule
6. Fachoberschule/Berufskolleg

- vocational school (part-time)
- higher secondary vocational school
- secondary modern school
- higher secondary school
- grammar school
- vocational school (full-time)

Secretarial activities

1 A company profile

Imeco International is an import/export business. Up until recently they have concentrated their business largely within Europe. But they have now decided to expand into America and the Far East. This is part of the homepage which they have produced for potential new clients:

Location: http://www.imeco.com

Welcome to IMECO International

Established in 1985, Imeco International is a dynamic and successful import/export company based in Dortmund at the heart of Germany's most important industrial region. In the last ten years we have built up a solid network of partnerships with firms throughout the European Union and the growing markets in Eastern Europe. Our business is expanding and we feel sure that both North America and the Far Eastern clients will also benefit from our unrivalled experience in the European market.

We specialize in the import and export of household goods and equipment. If you would like to find out more about us, please call:
Anke Schmidt in our Export Department on
++49 (0)231 149381-29 or email us at info@imeco.com

R **a** Answer the following questions.

1. What kind of business is IMECO involved in?
2. Where are their headquarters located?
3. When was the company founded?
4. Where are they currently doing most of their business?
5. What plans do they have for the future?

P **b** Work with a partner. Invent a company and draw up a company profile. Be ready to present your company to the class. → SF 13

Intercultural communication:
Types of business organisations
The British Private Limited Company (Ltd) and the American Closed Corporation (Corp.) are similar to the German GmbH. The German Aktiengesellschaft (AG) corresponds roughly to the British Public Limited Company (PLC or plc) and the American Stock Corporation (Inc. or Corp.). However, there are differences regarding liability, ownership and management.

2 Telephoning in business

An important part of the job of secretarial staff is to make and to answer telephone calls. This includes arranging appointments, making travel arrangements like booking a flight or making reservations at hotels and restaurants and of course taking messages and leaving messages on a voicemail. You may also be required to welcome customers.

a Describe the situations above.

b Choose the best phrase for each of the following situations.

1. You want the caller to wait.	Hang on, please. Hold the line, please.
2. You want to give your name on the phone.	This is … My name is …
3. You didn´t understand the other person.	I´m afraid I didn´t get that. I´m afraid I don´t understand that.
4. The person the caller wants to speak to is eating a burger.	I´m afraid he's eating a burger. I'm afraid he´s having lunch.
5. You need to check the name of the caller.	What's your name? Who's calling, please?

Intercultural communication: Politeness on the phone

When calling someone in an English-speaking country it is very important to use polite and suitable language. Don't say: "I want to speak to Mr Clarke" because this would sound impolite and unfriendly. It's more polite to say: "I would like to speak to Mr Clarke, please."

If you know the other person, you should start the conversation with a bit of small talk (e.g. How are you? How is the weather/the family? etc.).

In addition, people tend to say things in an indirect way. If you want the other person to do something for you, you could start with phrases like the following:

"It would be very kind of you …" / "Would you mind …". / "Could you possibly …"

Phrases for telephoning

To phone somebody	This is James Moore. Could I speak to Anke Schmidt, please?
	Could I speak to somebody in the sales department, please?
	May I speak to whoever is responsible for quality complaints?
To answer the phone	Good morning/afternoon. Imeco International. How can I help you?
	Who's calling? May I have your name please?
To say that somebody is not available	I'm sorry, but Anke Schmidt is not in the office at the moment.
	… is at lunch/in a meeting/on a business trip.
	… is off sick/… is not at her desk.
To offer to ring back or take a message	Can I take a message?
	Would you like to call back this afternoon or should Ms Schmidt call you?
	I'll ask Mrs Schmidt to call you back as soon as possible.
To get out of trouble	I'm sorry, the line is very bad/I didn't get/catch that. Could you repeat that, please?
	Could you speak more slowly, please?
To end a conversation	Goodbye and thanks a lot for your help.
	Goodbye Mr Miller. You'll be hearing from us again soon.

c Listen or read.
A prospective client has seen Imeco's homepage and is now phoning Anke Schmidt in the Export Department for more information. Her colleague Melanie Walter answers the phone.

MW: Guten Tag, Imeco – wie kann ich Ihnen behilflich sein?
JM: Hello, yes – do you speak English?
MW: Yes, sir. How can I help you?
JM: Well, this is James Moore of Seattle Trading in Seattle speaking. Could I speak to Anke Schmidt, please?
MW: Hold the line, please, I'll put you through. … I'm sorry, Mr Moore, Anke Schmidt is in a meeting just now. Can I take a message?
JM: That would be very kind. Could you tell her that Seattle Trading are very interested in the promotion material on your homepage? Perhaps she could call me back.
MW: Certainly, sir. Could I just have your name again and your telephone number, please?
JM: The name's Moore – James Moore, that's M-O-O-R-E on Seattle 296 54730.
MW: 296 54730 – right, thank you, sir. She'll call you back as soon as possible. Goodbye.

R **d** Now listen to another call to Imeco and write down the following information: name of the caller/where he is calling from/the name of his company/the reason for calling/his work number/his mobile number.

I **e** A third caller, Mr John Winter of Chicago Electrics (Chicago 312 884092), has also seen the homepage and wants to speak to Anke Schmidt but she is not available. → SF 16+18
With a partner, role play a conversation between John Winter and the receptionist. Use the phrases in the box above.

3 Appointments

Phrases for arranging appointments

To make appointments	I'd like to make an appointment to see Ms Schmidt, please. Would it be possible for me to see Ms Schmidt next week/on Tuesday? Would 10 am on Tuesday be convenient?
To accept and to reject suggestions	Yes, I can make it at 10 am on Tuesday. Wednesday would suit me fine. Unfortunately, I'm engaged all day on Tuesday. I'm afraid I've got a meeting then.
To change or cancel existing appointments	I'm afraid there's a problem. Can we postpone our appointment until 11 am on Wednesday? Ms Schmidt is ill. I'm very sorry, but we'll have to cancel your appointment on Tuesday.

Intercultural communication

The date can be written in various different ways: e.g. 3 January 2006 - 3rd January 2006 - January 3, 2006. Therefore it is advisable to write out the date when writing letters and making appointments. (Don't write: 03/01/2006 as this could be understood in America as 1 March 2006!)

a That afternoon Anke Schmidt returns John Winter's call.
Complete their conversation using expressions from the box on the right.

JW: Good afternoon, John Winter **(1)**.
AS: **(2)**. This is Anke Schmidt of Imeco International. I'm sorry that I missed your call but I was in a meeting all morning. **(3)**
JW: Hello, Ms Schmidt. Thank you for calling back. We have seen your homepage and I'd like to know more about your products. I'm planning to be in Germany in May and **(4)**
AS: Of course. **(5)**
JW: How about the 4th? **(6)**
AS: Just let me check my diary. Oh dear, I'm afraid **(7)** **(8)**
JW: Certainly, no problem. **(9)** Shall we say at ten o'clock?
AS: **(10)** I look forward to seeing you.
JW: Yes, so do I. **(11)** Goodbye.

- I'm free in the morning then.
- speaking
- I'd like to make an appointment to call in and see you.
- At about two o'clock?
- Good afternoon, Mr Winter.
- I'm already rather busy on the 4th.
- What can I do for you?
- What day did you have in mind?
- Fine - ten o'clock on the 5th of May.
- Could I suggest the 5th instead?
- Till the 5th then.

P **b** Draw a diary for Anke Schmidt and John Winter from Monday May 27th to Friday May 31st on a separate piece of paper. Then listen to the phone conversation following their meeting and fill in each speaker's appointments.

I **c** Now look at the two diaries on the right and role play a conversation in which Mr Toby Weismann of Weismann Associates in Boston makes an appointment with Anke Schmidt.

Toby Weismann	July
	12 Monday 14.50 arr. Munich 16.00 Expo / H. Bielstein
	13 Tuesday Dortmund 16.00 Hotel Hansa
	14 Wednesday

Anke Schmidt	Juli
	12 Montag 14.00 Abteilungskonferenz
	13 Dienstag 15.30 Mr Payne – New York Life Magazine
	14 Mittwoch 14.00 Ms Bone UK Services Ltd

I **d** Work with a partner and role play the following situation back to back. Make notes first and add names and polite forms.

A: Sie sind Mitarbeiter der Firma Ecotec GmbH in Mannheim. Ihr Verkaufsleiter Herr Schäfer hatte einen Termin für morgen 14.30 Uhr mit Tina Jones von der Firma Environment Ltd in Newcastle (GB) vereinbart. Leider ist Herr Schäfer krank. Sie bitten deshalb den Termin auf Mittwoch nächste Woche zu verlegen.

B: Sie sind Assistent von Ms Jones und antworten, dass dies möglich ist. Der Termin muss jedoch nachmittags stattfinden, da Ms Jones am Vormittag eine Besprechung hat.

P **e** You work for the German firm Classic Cars Import, Am Schlosspark 15, 60485 Frankfurt/Main. Your boss, Michael Moser, has asked you to write a business letter (date: 15th March) to MG Cars, 24 Durbon Road, Laleham, TW15 1SD, UK, to make arrangements for his visit. He gave you the notes on the right:

- komme am 25. April um 15.00 Uhr an
- bleibe bis zum 27. nachmittags
- werde im Hotel Mercury wohnen
- möchte den Verkaufsleiter von MG Cars am ersten Abend im Hotel treffen
- Details zum Import von Oldtimern sollen besprochen werden

4 Booking flights and trains

Jochen Westkamp, his secretary and two of his assistants are in America on business. At the moment they are staying in New York but have decided to extend their trip to Dallas. Jochen has asked his secretary, Greta Fabian, to arrange their flight. Here is her telephone conversation with a local travel agent (TA).

TA: New York City Travel Bureau, can I help you?

GF: Yes. This is Greta Fabian of Imeco. I'd like to book a flight from New York to Dallas on Wednesday of this week for four people.

TA: No problem. Shall I check what's available in business class or will you be travelling first class?

GF: Business class, please.

TA: Well now, business class to Dallas on June 3rd – there's a flight with Delta Air Lines at 16.30, that's non-stop, or an American Airlines flight at 13.30 with one stop in Philadelphia. Seats are still available on both.

GF: What time does the 16.30 flight arrive?

TA: It arrives at 19.35 local time.

GF: At 19.35 – and the 13.30 flight?

TA: You have an hour in Philadelphia with that one, and you get into Dallas at – let me see – 17.35.

GF: Well, I think we'd better take the one at 13.30. That will leave us enough time to get to the hotel.

TA: Right, so is that a firm reservation?

GF: Yes, it is, thank you.

TA: Good. Could I just take your name and credit card details and I'll have the tickets ready for you in the morning.

GF: Certainly. It's Greta Fabian – that's F-A-B-I-A-N, and the credit card number is 4929 824 752 311, Barkers Bank.

TA: Thank you very much. Goodbye now.

a Complete the two conversations using suitable words or phrases from the dialogue above.

1. - I'd like to **(1)** a seat on flight AA 496 to Los Angeles in **(2)** class on June 13th, please.
   - I'm sorry, madam. There are no seats **(3)** on that **(4)**. Shall I try a later one?
   - Yes, please.
   - There's a flight at 16.30 which arrives at 17.10 **(5)**.
   - Yes, that should be OK.
   - Shall I make a **(6)** then?
   - Yes, thank you very much.
   - Could you give me your name and **(7)** number, please?

Phrases for booking flights and trains

To ask about flights or trains
What time would you like to arrive in Munich?
Do you prefer a certain airline?
Are you travelling business or economy class?
Is that first or second class, one way or return?
Would you prefer a non-smoking/smoking compartment?
I need to be in Munich at 2 pm at the latest.
Please reserve an aisle seat on the 9.30 am flight to Munich.

To give instructions about flights or trains
Please check in two hours before departure.
You are booked on flight LH 431 on 24 May, departing London Heathrow at 9.30 am, arriving in Munich 12.30 pm.
I've reserved a window seat on the 10.00 am ICE train to Hamburg.
You can pick up your ticket at the LH desk at the airport.

I **b** In Dallas Jochen Westkamp and his colleagues decide to fly on to San Francisco. Greta phones the travel agent, who looks at the Dallas airport flight schedule. (Day 1 = Monday, etc.)
With a partner, take the roles of Greta Fabian and the travel agent.
Greta wants tickets in business class. She wants to know what days they can fly, the arrival and departure times and whether there are any stops on route. → SF 16+18

	DALLAS		SAN FRANCISCO	
Days	**Depart.**	**Arr.**	**Airline & Flight No.**	**Stops**
1-2-4-6	8.30	10.20	DL 1291	non-stop
1-2-4-6	12.30	14.20	DL 1492	non-stop
3-5-7	10.40	13.10	AA 862	Denver
3-5-7	17.20	19.50	AA 894	Denver

P I **c** Back in Germany Greta finds this note from her boss on her desk. Read the note and draw up a telephone dialogue together with your partner. Invent a problem and find a solution. Then role-play the situation back to back.

Greta
Bitte buche eine Zugfahrt erster Klasse von Manchester nach Leeds für den 16. Mai. Habe dort um 16.00 Uhr einen Termin. Rückfahrt am folgenden Tag morgens so gegen 10.00 Uhr.
Vielen Dank
Jochen Westkamp

5 Reservations at hotels and restaurants

In Dallas Greta Fabian also has to make a hotel booking for herself and her three colleagues to stay in San Francisco. One of Jochen's business associates has recommended a good hotel and given her the internet address of the homepage below.

a Here is the phone call which Greta makes to reserve rooms. Match what she says with what the receptionist at the hotel says and put the dialogue in a logical order.

R: Golden Gate Hotel. Can I help you?

GF: Hello. Greta Fabian of Imeco speaking. I'd like to book four rooms for five nights from the 7th of June, please …

Now you go on!

Receptionist

1. Yes, no problem. We'll hear from you later then.
2. Shall I reserve the rooms for you then?
3. We have four singles available, but only two with bath. The other two are with shower – is that all right?
4. I'm afraid not, madam. It'll be $15 extra for breakfast.
5. Anything else I can do for you?
6. Just hold the line, I'll check for you. Would you like single or double rooms?
7. We have a small conference room for twelve people available at $80 per day.

Greta

a) Singles please, with bath if possible.
b) Oh, I'll check that with my boss first and confirm the booking later today. Will that be OK?
c) Yes. I've seen your homepage, but I'm not sure if the price includes breakfast.
d) Yes, that'll be fine.
e) So that's $115 per night with breakfast, OK. Now we'd also like a small conference room for those four days.
f) Oh, that's about right, we're only about ten people.

P **b** Greta has checked with her boss and rings back later in the afternoon. Write an e-mail to the Golden Gate Hotel including all necessary details from the dialogue. Ask for a written confirmation.

I **c** Jochen's business associate has also recommended a good restaurant nearby. Greta telephones the restaurant to reserve a quiet table for herself, her three colleagues and two business partners for 8 pm on 8 June. Draw up a telephone dialogue together with your partner and then role play the situation back to back. Use the phrases in the box.
→ SF 16–18

Phrases for making reservations at hotels and restaurants

To book hotel rooms and facilities
Could you reserve a single room with bath for the two nights 10th and 11th September?
I should like to reserve a superior room with internet access and twin beds/a double bed from 2 to 5 April.
We require a conference room to accommodate at least 15 people.
The conference room should be equipped with an overhead projector/computer and video projector.
Would you mind confirming the booking in writing/by fax/by email.
Could you send the bill to us?

To make eating arrangements
I´d like to book a quiet table for 3 people for 8 pm on Tuesday, please.
Is it possible to get a table at the window?
What time do you start serving lunch/dinner.

To change existing arrangements
We regret but we have to cancel our reservation.
Unfortunately we need to change our reservation of 16 April. We need the rooms for three nights instead of two nights.

6 Netiquette

Netiquette is network etiquette and means a set of rules for behaving properly online. The following rules of e-mail Netiquette will help to communicate over the Internet.

Match the beginnings of the rules with the corresponding endings.

1. Be brief and	a) don't change the original wording.
2. Mails should have a subject title that	b) capitalizing looks as if you are SHOUTING.
3. Check with your employer	c) when using sarcasm and humour.
4. Don't send anything in an e-mail	d) unwelcome and forbidden in many countries.
5. If you forward a message,	e) that you wouldn't write on a postcard.
6. Never send chain letter	f) don't send any larger than about 50k.
7. Use mixed case,	g) through the Internet.
8. Unsolicited e-mail advertising is	h) keep your messages to the point.
9. Be careful	i) reflects the content of the message.
10. When attaching files,	j) if private e-mails are allowed at work.

7 Invitations

Jochen Westkamp of the Purchasing Department at Imeco sends this fax to invite a business partner – Mr Martin Hope, the European representative of Singapore Trading – to Dortmund to present his new products.

Intercultural communication: Politeness
When you write a business letter, it's important to use a tone that is very friendly and polite. Being factual und to the point like in German correspondence might seem rude in the English-speaking world. Therefore, you should use expressions like "We would be very grateful …" or "Would you please be so kind …" for requests and "We are extremely sorry …" or "I'm afraid …" for problems and difficulties. Do also not forget to use the word "please" whenever appropriate.

Telefax

From/Von:	Jochen Westkamp, Imeco International
To/An:	Martin Hope, Singapore Trading Consultants
Date/Datum:	16 July …
Subject/Betreff:	Invitation to meeting in Dortmund

Dear Mr Hope

I am writing to let you know about a meeting of our representatives which we have planned for 28-30th September here in Dortmund. **We would like to invite you (1)** here on one of those days in order **to present your new range (2)**. All our major representatives will be here and **I feel sure that (3)** it would be of great mutual benefit if you were to meet them as soon as possible. **Could you let me know whether you will be able to attend? (4) I do hope you can make it (5)**.

Yours sincerely
Jochen Westkamp

a Replace each of the numbered expressions above with two similar ones from the box below.

1. to show us your new products
2. let you have the particulars
3. When I have received your reply
4. I would like to ask you to join us
5. I am confident
6. to display the new items in your collection
7. fax a provisional agenda
8. I wondered if you would also like to come
9. doubtless
10. Looking forward to your positive reply
11. I would be grateful if you could tell me whether you will be coming.
12. It would be nice if you could come

P **b** Using these notes which Martin Hope gave to his secretary, write a fax accepting the invitation.
Dear Mr Westkamp / many thanks for invitation to … / delighted to accept / as you say great mutual benefit to meet your representatives / be a pleasure to see you again / best wishes … → SF 11

8 Welcoming a guest

When Martin Hope of Singapore Trading arrives at Imeco's offices in Dortmund, Greta Fabian meets him for the first time.

I **a** Complete the following conversation with words from the box and then role play the situation back to back.

pleasant · bumpy · refreshments · great · bright · welcome · too · I'm sorry – afraid · help yourself

Greta:	Did you have a **(1)** flight?
Mr Hope:	I'm **(2)** not. The flight was rather **(3)**
Greta:	Oh, **(4)** to hear that. If you'd like to come this way, Mr Westkamp is waiting for you in our conference room. … here we are.
Mr Westkamp:	Hello Mr Hope. **(5)** to our company.
Mr Hope:	Nice to meet you, Mr Westkamp.
Mr Westkamp:	Nice to meet you, **(6)**.
Greta:	Can I offer you some **(7)**?
Mr Hope:	A cup of coffee would be **(8)**.
Greta:	Here you are. Please **(9)** to milk and sugar.
Mr Hope:	Thank you. By the way, I'm really impressed by your offices. They are very **(10)** and modern.
Mr Westkamp:	Well, yes, the offices and the furniture …

b Match the questions with the answers.

1. Did you have a pleasant flight?	a. No, not really. I'm more into tennis.
2. Are you interested in soccer?	b. Thanks, but I would like to freshen up first.
3. How do you like your hotel?	c. Not really, the flight was rather bumpy.
4. May I take your coat?	d. Yes, thank you.
5. Would you like to take a chair?	e. No, actually the flight was 30 minutes late.
6. What was the weather like in France?	f. It's great. Very quiet and very friendly staff.
7. Was the flight on time?	g. Just sugar, please.
8. Do you take milk and sugar?	h. It was cold and lots of rain.

9 Giving directions

■ John Winter has arrived at Imeco's office building in Dortmund to meet Anke Schmidt. He is standing at the reception desk. The receptionist directs him to Ms Schmidt's office (Export Assistant). Take the role of the receptionist. Use the floorplan and the phrases on the right.

Phrases
Go up/down the stairs.
Take the lift to the first/second floor.
Go across the hall.
Ms/Mr … office is on your left/right.

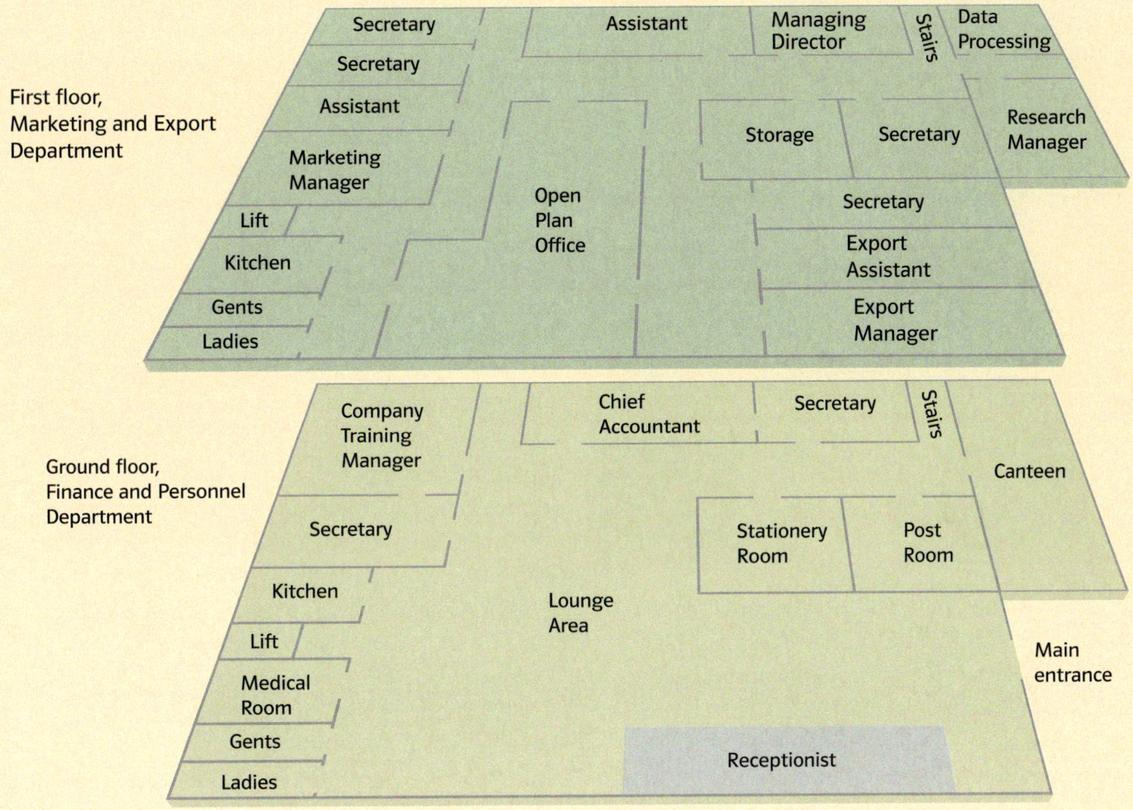

Intercultural communication: Ground floor (GB) – First floor (US)
When giving directions in a building it's important to remember that the British call the German "Erdgeschoss" ground floor and the "1. Etage" first floor whereas the Americans call it first floor and second floor.

Unit

15 Poverty and the welfare state

Growing up on welfare

The stereotype of the American poor is "black and lazy". That is why many middle-class Americans believe that their tax-money is being wasted on welfare payments. They insist that a lot of these payments go to lazy black women who are only interested in having sex without birth control and also do not want to get married. Rosemary L. Bray, a black woman from Illinois, tries to explain what living on welfare really means by telling us her own childhood experiences.

"I know what it means to be poor. And I know that the welfare system leaves people on the edge of survival. I know because I have been there. My parents came to Chicago in 1980. My mother got a part-time job in a restaurant. Father worked at whatever came to hand – from selling junk at open air markets to selling hamburgers on the street. When I was born, he made my mother quit her job. They had, like most American couples, their own American dream – a husband who goes to work, a wife at home and a family of smiling children. But, as it is true for so many African-American couples, their dream was an illusion. After being married for several years my mother realized that my father was a gambler and an alcoholic. One day he didn't return home and he has not been back since. After discussing her case with the social workers, she decided to sign up for AFDC*. The welfare payments for the rent and for all other expenses, by any standards, were not really sufficient. There were times when we hardly had a loaf of bread. So what did it take to survive? It took the kindness of friends and strangers, the charity of the churches, low expectations, deprivation and a lot of patience.

What really made our lives work was my mother's genius at "making do". Shopping for food was done at the cheapest stores. Clothes were bought at charity second hand stores or at street markets. However Mama's biggest worry was that we children should grow up uneducated. So she sent us to a church-run private school. When one social worker asked her how she could afford to send us there, she declared "My kids need an education" and told him to mind his own business. She reminded us again and again that every book, every test, every page of homework was in fact a ticket out and away from the life we were leading at that time.

We lived on welfare for 18 years. In the meantime Mama has died and we children are all tax-paying working adults. When I read articles about the abuse of the welfare system in America I get angry. Most of them are full of lies and prejudices. Of course, some women are lazy and misuse the system. But the vast majority of welfare recipients are honest and hard-working. And life on welfare is no picnic.

496 words

** AFDC = Aid for Families with Dependent Children*

R a Multiple choice

Find the most suitable answer.

1. Which prejudice is mentioned in the text ?
   A most black people are lazy
   B most black people are dirty
   C most black people do not get married
   D most black people are criminal
2. Why was the American dream an illusion for the Brays ?
   A because they were immigrants
   B because they were poor
   C because their family broke up
   D because the children couldn´t go to school
3. How did the family manage on welfare payments ?
   A with the help of friends
   B by stealing things they needed
   C by begging strangers for help
   D by buying cheap clothes and food

R b Reading comprehension

Give a short answer from the information you find in the text. → SF 1

1. How does Rosemary explain what living on welfare really means?
2. What did Rosemary's father sell?
3. What had Rosemary's father become?
4. What were the AFDC payments for?
5. Where did Rosemary`s mother send her children?
6. What did she believe education was for her children?

c Gapped summary

Complete the following sentences with words from the text "Growing up on welfare".

Rosemary points out that they lived on … **(1)** … for almost twenty years. Now they are all … **(2)** … and they have jobs and pay … **(3)** …. Rosemary adds that she feels very … **(4)** … when she reads … **(5)** … about the abuse of the welfare system by black people. Of course, some black people … **(6)** … the welfare system, but they are not the … **(7)** …. Most of them are … **(8)** … and work hard. Such articles are filled up with … **(9)** … and … **(10)** … .

M d Translation

Translate lines 51 to 59. → SF 10

P e Composition

Comment on the following statement, using your own knowledge and ideas.

"Welfare payments should be kept low so that recipients adopt more responsibility for their own lives." → SF 5

1 Listening comprehension: Poverty and young people

a Listen to what these three young people have to say about their lives and then answer the following questions in short answers. R

1. Where do the young people come from?
2. How old are they?
3. What problems do they have?

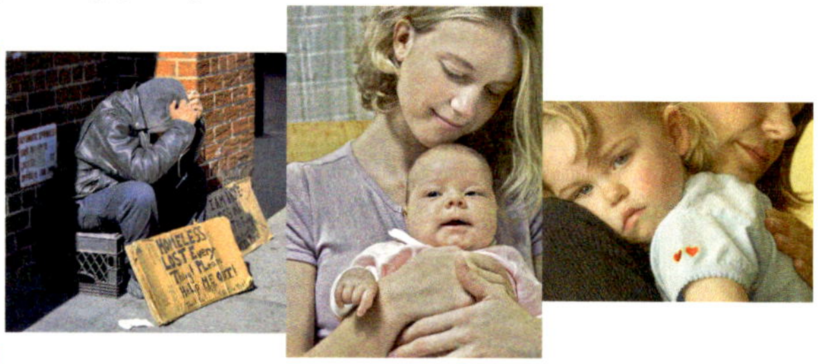

b Listen again and find the most suitable answer. R

1. Sarah would like to
   a) stay at home.
   b) go out in the evenings.
   c) get a job.
   d) get married.

2. Sarah says her problem was her fault because she
   a) should have been more careful.
   b) shouldn't have gone to the discotheque.
   c) should have had more money.
   d) should have got help.

3. Steve got a job training
   a) in a factory.
   b) in a centre for the homeless.
   c) at a garage.
   d) at a hotel.

4. Steve is living on the streets now because
   a) he lost his job.
   b) his friends do that.
   c) he likes it.
   d) his parents threw him out.

5. Mandy's boyfriend left her because he
   a) had no money.
   b) loved another girl.
   c) wanted to be free.
   d) had a job training in another city.

6. Mandy gave up her job training as a nurse because she
   a) didn't like the job.
   b) moved to another city.
   c) got pregnant again.
   d) had to look after her baby.

2 Describing statistics: Child poverty → SF 21

P ■ Describe and analyse the following diagrams showing child poverty in the USA.

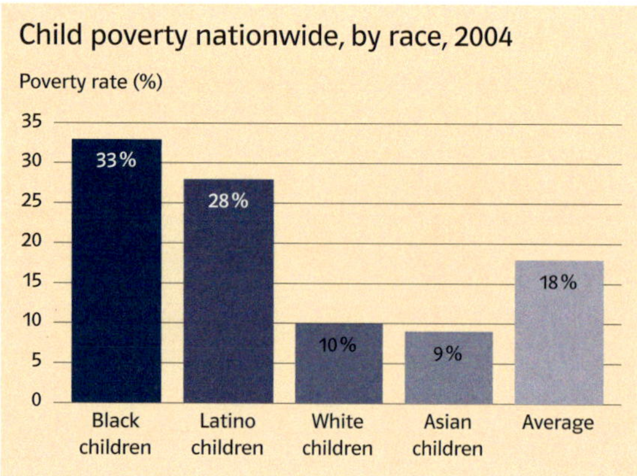

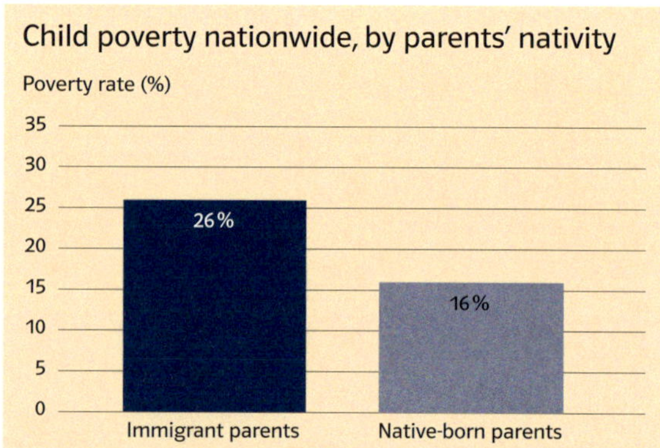

((Quelle: National Center for Children in Poverty – www.nccp.org/pub_cptb.html))

1 The cycle of deprivation and poverty

a After you have studied the diagram below, complete it with suitable words from the box.

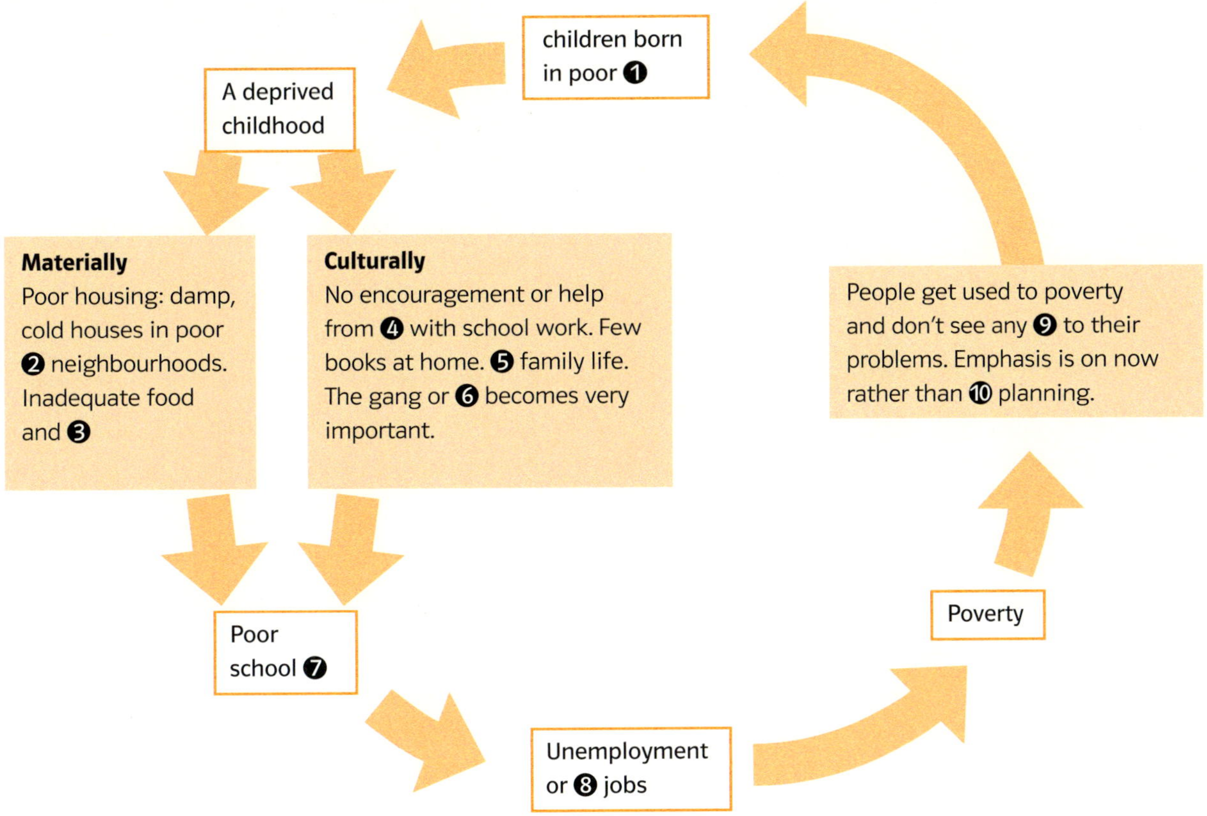

inner-city – clothing – parents – peer group – solutions – families – future – marks – long-term – inferiority – husbands – educational – wishes – poorly-paid – unstable

b Now describe the diagram in your own words.

2 Voluntary organisations

P First describe the cartoon on the right. Then discuss the problem which is shown in it. → SF 20

Unit

16 Juvenile delinquency

"It wasn't like that in my day"

Recent newspaper reports have been full of complaints about the declining standards of teenage behaviour. A lot of people say that young people are suffering from a loss of values. However, according to a study by Leslie Francis from the University of Wales, many young people are just confused about what is right or wrong. A survey of 14,000 schoolchildren aged 13 to 15 found that 17 % believe there is nothing wrong in playing truant from school. 16 % think that travelling on public transport without a ticket is not wrong; 6 % feel there is nothing wrong with shoplifting; and only 29 % said that they had never stolen anything. Although ideas about right and wrong or good or bad depend on people's viewpoint, sociologists can judge a society's morals by using crime levels as indicators. They can also interview people to find out their opinions on issues such as smoking, drinking and drug taking.

But was it better in the past? Were young people less immoral and less violent than they are today? According to Dr Henry Henrick, a British sociologist at Oxford University, young people tend to develop the values of the society in which they grow up. Moreover, links between social poverty and crime are well known. One survey of Newcastle's young people found that 60 % of boys and 9 % of girls from "deprived families" eventually became criminal. This survey suggests that poverty has a negative effect on the parent's ability to educate their children in a socially acceptable way. Another factor which leads to juvenile delinquency is unemployment. The British Youth Council has recently published figures which show that unemployment is running at 16 % for the under-25s. Income support for most 16- and 17-year-olds has been abolished and the government's promise to provide youth training places for every teenager remains unfulfilled. What is worse, long-term unemployment is linked with suicide. In the last ten years the suicide rate among the under-25s has risen by 30 %. Alcohol and drug taking is also growing. It has been proven that the typical young person who has been unemployed for a longer period of time is likely to fall victim to a wide variety of accidents and crimes.

In his book "Hooligans: a report of respectable fears", Geoffrey Pearson argues that adults have always had fears and prejudices with the so-called young rebels. According to him, because society always looks at its past days through rose-tinted glasses, it suffers from historical amnesia. In other words, when adults look at young people's morals, they always maintain that they were better in the good old days.

Such forgetfulness about past times can lead to moral panic in which adults are deeply shocked by children's behaviour and as a result start to blame individuals instead of looking to wider causes.

Young people will always create their own style, but do their morals and behaviour really change from one generation to the next? It is the job of sociologists to analyse situations and to suggest help. All of us, as members of society, have then the duty to try and find possible solutions to the problems which young people face.

532 words

(The Guardian, adapted)

a **Reading comprehension** → SF 1

1. Look at the photo and say what you expect the text to be about.
2. What do you think the headline has to do with the photo?
3. Now read the text quickly to find out if your ideas were right.

R **b** **Multiple choice**

Find the most suitable option.

1. Which crime can you find in the text?
   a) shoplifting
   b) murder
   c) playing truant
   d) vandalism
2. How can sociologists find out what people believe is right or wrong?
   a) by studying crime rates
   b) by going to church
   c) by interviewing people
   d) by studying sociology books
3. How are crime and poverty connected?
   a) when there are more boys and girls in large families
   b) when parents do not bring up their children well
   c) when people are out of work for a long time
   d) when teenagers leave home
4. How does long-term unemployment affect young people?
   a) the percentage of young people who kill themselves rises
   b) more youngsters become delinquent
   c) they decide to leave home
   d) more young people become hooligans

M **c** **Mediation**

Übertragen Sie die Zeilen 49 bis 69 ins Deutsche. Übersetzen Sie nicht Wort für Wort. → SF 10

d **Gapped summary**

Complete the following sentences with words from the text on page 118.

In the last few weeks we can find a lot of … **(1)** … in newspapers about young people's behaviour. According to some adults, teenagers … **(2)** … from not knowing what is right and what is … **(3)** …. In a … **(4)** … thousands of schoolchildren were interviewed. Almost a fifth of the children interviewed said there was nothing bad in … **(5)** … … **(6)** … from school. Almost a third of the young people who were interviewed admitted that they had never … **(7)** … anything up till now.
This seems to underline the opinions that a lot of … **(8)** … have of children - and that standard of teenage behaviour is … **(9)** ….

e **Composition** → SF 5

P Comment on the following statement, using your own knowledge and ideas.
"Young people today do not know the difference between what is right or what is wrong."

1 Listening comprehension: Young offenders

R ■ Listen and find out if the following statements are true or not true. Correct the false statements.

1. Mark
   a) Mark steals because he needs the money.
   b) Mark's criminal career started with shoplifting and smashing bottles.
   c) Mark's friends were into stealing cars.
   d) Mark realises that stealing is wrong.
   e) Mark wants to work hard so that he will be rich in the future.
2. Jenny
   a) There are only drug takers and drinkers in Jenny's neighbourhood.
   b) 80 % of the adults that Jenny knows take drugs.
   c) Jenny has been taking drugs all her life.
   d) Jenny would prefer a normal life without drugs.
   e) Jenny sees herself as a criminal.

2 Describing a cartoon: School violence → SF 20

P ■ Describe what you see in the following cartoon and try to explain its message.

1 A case study: Death on the streets

a Describe what you can see in the photos. → SF 19

b What do you think was the reason for this car crash?

c Now read the following text to find out the reason.

In this interview a young boy (15 years old) says why he started joyriding:
"When I was 14 I got a Saturday morning job at a scrap yard. I started playing about with old cars. I knew how to drive because I had watched my dad. Well, one Saturday evening we didn't have much to do – there's never much to do in our town. My mates suggested going on a joyride, just for a bit of fun. They knew I could drive a bit. So we broke into this car and off we went down the motorway. My mates told me to drive faster and faster. It's difficult to describe the feeling you get. I suppose it's about being in control of so much power. Just like in the movies or on TV adverts. I didn't really think about the risk we were taking."

I **d** Discuss this case study in class. In your discussion you should try to find reasons for and solutions to the problem. → SF 14

1. Role play:
   - Student A: a young joyrider
   - Student B: a social worker
   - Student C: a teacher
   - Student D: a parent

Act out a discussion between these people.

P **e** Write a report on the results of this discussion. → SF 7+8

2 Describing statistics: Stolen cars

P **a** Describe the following graphs. → SF 21

Why cars are stolen

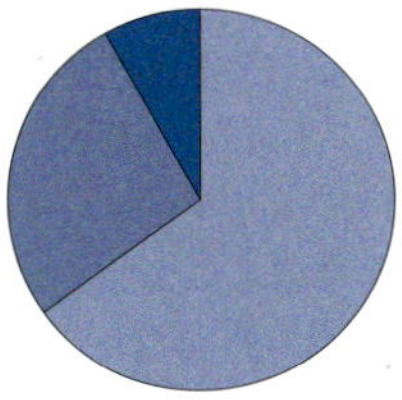

- 65 % Theft for temporary personal use (joyriding)
- 8 % Insurance fraud
- 27 % "Professional" theft (for financial gain)

Who steals cars?

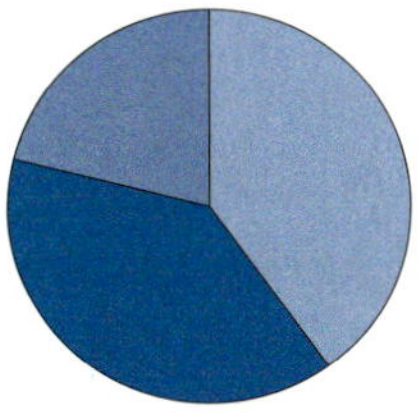

- 40 % between 17 and 21 years old
- 39 % under 17 years old
- 21 % Others

- 83 % male
- 17 % female

b Try to explain the figures.

A future for the earth?

"One generation passes away, and another generation comes: but the earth lives on forever." (Holy Bible)

No, not forever. For man has now reached a point in his evolution where he has the power to influence the present and the future state of the planet, for better or for worse.
Since the beginning of the Industrial Revolution, smokestacks have released harmful gases into the atmosphere, factories have dumped toxic wastes into rivers and streams, cars have consumed irreplaceable fossil fuels and polluted the air with their exhaust gases. In the name of progress, forests have been cut down and lakes and oceans poisoned with pesticides and chemicals.
The continuing destruction of the atmosphere's ozone layer is partly caused by the use of atmosphere-destroying chlorofluorocarbons (CFCs) in different types of foam, insulation and air-conditioners. Perhaps most threatening of all, the destruction of the tropical forests, home to at least half the earth's plant and animal species, continues at a rate equal to one football field a second. The resulting worldwide extinction of species promises to be at least as great as the mass extinction that happened at the end of the age of dinosaurs.
For a long time scientists have pointed to the damage and its possible consequences. But no one has paid much attention. Now things have become so bad that people are beginning to listen. Surely everyone must now notice that the Earth is in danger. In many countries heat waves every summer now drive temperatures above 40 degrees Centigrade, raising fears that the "greenhouse effect" – global warming – is already under way. Because of the lack of rain, some forests, including the Yellowstone National Park in the U.S., have already gone up in flames several times. And on many beaches, garbage, raw sewage and medical waste spoil the fun of bathers and confront them personally with the growing pollution of the oceans. Pollution in Europe has closed beaches on the Mediterranean, the North Sea and the English Channel. Thousands of seals have died because their immune systems have been weakened by pollution. What will happen, if nothing is done? The increase of CO_2 in the atmosphere could raise the planet's average temperature by 2 to 6 degrees Centigrade within the next fifty years. That could cause the oceans to rise by several metres, flooding coastal areas and ruining huge parts of farmland. Changing weather conditions can make large areas infertile and uninhabitable, resulting in migration on a scale which we have not experienced up till now.
Let there be no illusion. Although several world conferences (Earth Summit: Kyoto 1997, Buenos Aires 1998, Bonn 2001, Johannesburg 2002), have taken place to discuss global warming, and despite the fact that many companies have already been able to cut their emissions of carbon dioxide considerably, a lot more must be done. Taking effective action to halt the massive damage to the earth's environment will require a mobilization of political will, international cooperation and sacrifice unknown except in wartime. But humanity is at war right now, and it is no exaggeration to call it a war for survival. It is a war in which all nations must be allies. Both the causes and effects of the problems that threaten the earth are global, and they must be attacked globally.
Every individual on the planet must be made aware of the serious situation and of the need to do something. No attempt to protect the environment will be successful in the long run unless ordinary people – the American housewife, the Chinese factory worker, the African farmer and the European office worker – are willing to change their lifestyles.

586 words

R a Multiple choice

Find the most suitable option.

1. The text is about …
   a) pollution and its effects on life on our planet.
   b) the development of pollution in America.
   c) the future of the earth.
   d) the continuing destruction of our planet.
2. The photo illustrates …
   a) pollution.
   b) the earth's environment.
   c) the ozone layer.
   d) the greenhouse effect.
3. The word "their" on line 9 refers to …
   a) fossil fuels. b) factories.
   c) smokestacks. d) cars.
4. "Mass extinction" on line 22 means that …
   a) animals are killed.
   b) plants and animals die out.
   c) a lot of different plants and animals die out.
   d) no species are left.
5. An "infertile" area on line 48 is a place where nothing can …
   a) live. b) grow. c) survive. d) be healthy.
6. We are "at war" (line 61) means that we must …
   a) all try very hard in order to save the planet for future generations.
   b) change our lifestyles and protect the environment.
   c) mobilize our politicians and cooperate.
   d) do more than just discuss global warming.
7. The bible is quoted in the headline "One generation passes away …" in order to point out that …
   a) these problems also existed thousands of years ago.
   b) the situation has changed dramatically.
   c) the author believes in God.
   d) man has now replaced God.
8. The author has written this text because he wants …
   a) to describe the situation.
   b) to explain the situation.
   c) to give reasons for the situation.
   d) to change the situation.

R b Multiple matching

Match the effects with the reasons mentioned according to the text. There are two effects that do not fit.

Reasons	Effects
1. pollution	A. woods burn
2. lack of rain	B. the immune system is weakened
3. destruction of tropical forests	C. people have no clean drinking water
4. CO_2	D. animals and plants die out
5. rising temperature	E. millions of people leave their homes
6. factories	F. limited fossil fuels are consumed
7. changing weather conditions	G. the ice round the poles melts
8. traffic	H. temperatures rise
	I. the water is poisoned
	J. areas near the sea are flooded

R c Reading comprehension → SF 1

Answer the following questions in your own words as far as possible.

1. How has man, in the name of progress, changed the earth's environment?
2. How has nature reacted to such treatment?
3. What will happen if nothing is done about this serious situation?
4. Why does the author say that humanity is at war right now?

P d Letter to the editor

Write a letter to the editor of the magazine which published this article.
Give your ideas on the measures that governments, industry and ordinary people can take in order to improve the environment.

1 Listening comprehension: Recycling

R **a** Listen and find out which of the following materials are mentioned:

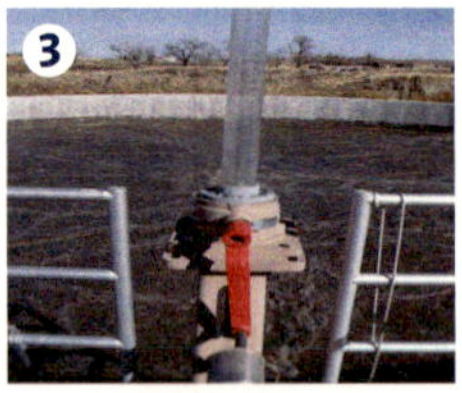

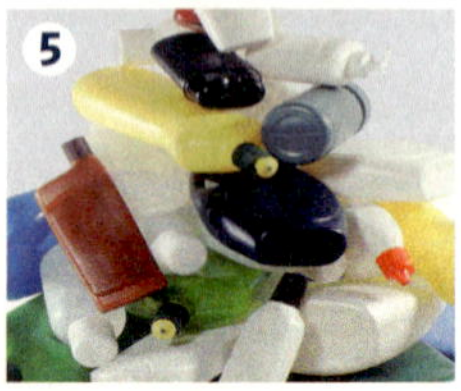

b Listen again and find the most suitable option. R

1. What would be the best title for this lecture?
   A. Important Keys to Recycling Paper
   B. Technological Advances Improve Recycling
   C. Steps to Improving Recycling
   D. No More Pollution

2. According to the article, paper materials that are difficult to recycle include:
   A. copy paper
   B. document shred
   C. food wrappers
   D. newspapers

3. Which fact is NOT mentioned?
   According to the lecture, the government could help companies which are involved in recycling by
   A. lowering taxes for these companies
   B. lending money at low interest rates
   C. giving money directly to them
   D. by participating in such companies

4. According to the article, the demand for recyclable materials in the manufacturing of new products is sometimes sluggish because
   A. some governments are unwilling to support expensive recycling methods.
   B. there is a lack of advanced technology to process the materials.
   C. businesses do not invest enough money into research.
   D. there is a lack of resources which can be recycled.

5. Which is NOT one of the main keys to recycling as mentioned in the lecture?
   A. government regulation of waste
   B. better technology
   C. more demand for recycled materials
   D. better informed consumers

2 Describing a cartoon: Pollution

P ■ Describe what you see in the cartoon and try to explain its message. → SF 20

Waste management

Millions of electric and electronic appliances are produced every year. But they have a short life. TV-sets are often replaced after seven or eight years, personal computers are technically out-of-date soon after they have been bought. So what to do with the rubbish?
Today more and more firms are trying to avoid or substitute harmful materials and to reduce the amount of scrap.

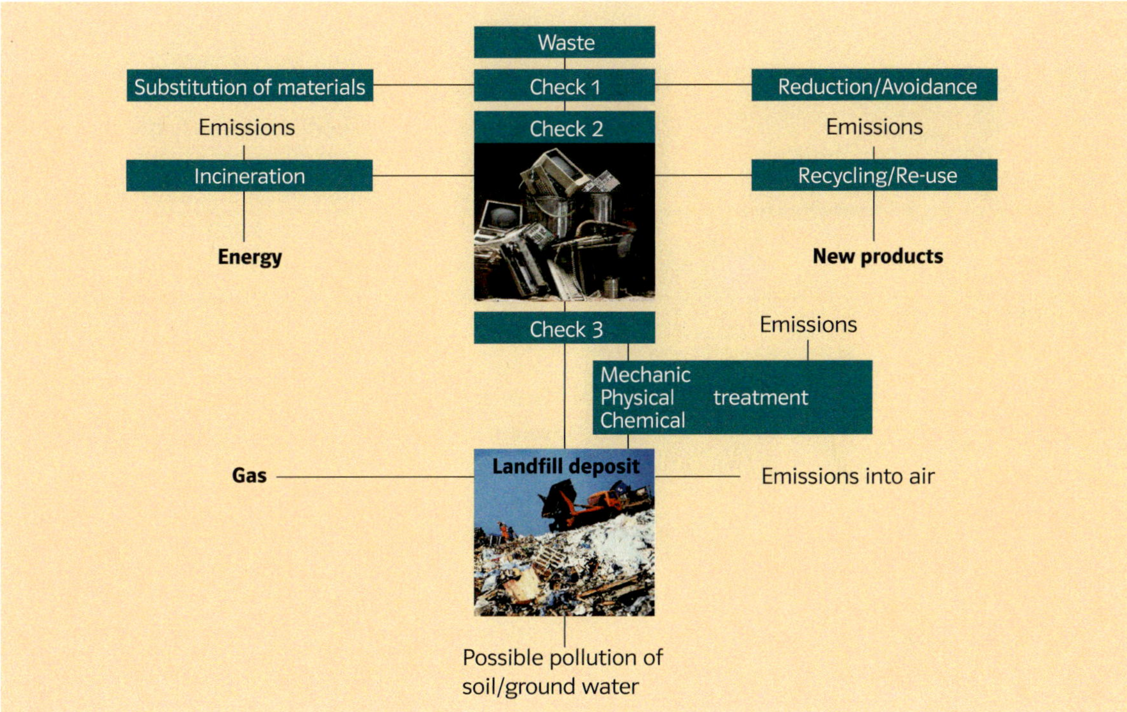

a Describe the decisions you should make at the different checks in the diagram above.
Use the words in the box.

clean · recharge · maintain · burn · avoid · substitute · reduce · treat · repair · melt · re-use · recycle · dump

Example:
At check 1 you must decide if you can substitute materials or …
At check 2 you …

b What would you do with the following parts of electronic appliances?
Think of reducing, substituting, recycling and incinerating.

1. plastic parts
2. paint
3. cardboard packaging
4. oil
5. batteries
6. mixture of plastic, metal, glass

c Give reasons for and possible disadvantages to your suggestions.

Unit 18 The internet

Bid for eBay

(A) Everyone has a favourite story about eBay, the online auction site. The less well-known story about eBay is its huge business success. Its steadily rising share price didn't even stop during the great dot.com crisis. Not long ago it broke the $60 barrier on the stock exchange.

By now eBay has more than 90 million registered users across the world, including more than 2 million in the UK and the company arranged sales of more than $2bn last calendar year. In the US analysts expect an annual growth of about 30%. So what's the secret? Certainly eBay has become much more than just a business name. What distinguishes eBay from most dot.com companies is that it was built on a solid business basis. Turning the internet into a global marketplace for new, used or rare goods was a simple concept that created maximum interaction of supply and demand.

(B) The only thing it needed was the participation of people – lots and lots of people. When times were good, people turned to eBay because it was a new way of doing business. When times turned tougher, eBay became the place to hunt for bargains on everything from old books at 50p to veteran cars at £20,000, from digital cameras to precious watches and a good place, too, to raise money by selling your own old things. It also created jobs – about 10,000 people make their living by selling goods on eBay.

(C) Not everyone has been surprised by eBay's success. The San Francisco economist Kent Sims observed recently that only overestimated dot.com companies didn't survive. The survivors such as Google, Yahoo!, Amazon, Schwab's online banking business or Diller's USA Interactive, which includes home shopping, travel or online concert booking, are based on a successful business concept.

(D) The whole thing was the idea of Pierre Omidyar, a French graduate student who designed a website from his Silicon Valley bedroom and offered a broken laser printer for sale in 1995. Someone bid $15 for it, and a new business – called Auction Web in the beginning – was born. From the start, Mr Omidyar was convinced that the key to eBay's success was people's willingness to trust each other. Now however, the company offers all sorts of safeguards including customers' feedback on sellers and a secure credit card payment system.

(E) However, cyber crime is never far away. Not long ago, fraudsters stole credit card information by creating a website that looked like eBay's. Former users were told that there had been problems with their payment and they were asked to enter their credit card details again. Moreover, dozens of chat rooms exist where thieves and hackers exchange personal details of card holders.

(F) So there must be a certain myth about eBay. "In such uncertain times, eBay continues to grow because about 100 million people trust a complete stranger," Omidyar said recently. "To me, it's proof that goodwill exists in all of us. I'm convinced that our common interests are more powerful than our differences."

500 words

R **a** **Multiple choice**

Find the most suitable option.

1. Just a little while ago eBay shares
   a) fell below $ 60.
   b) rose by $ 60.
   c) doubled.
   d) rose to more than $ 60.

2. More than 10,000 people
   a) work for eBay.
   b) work for eBay and its subsidiaries.
   c) depend on the online auction site as a main source of income.
   d) buy and sell on eBay every day.

3. Internet fraud
   a) doesn't affect eBay.
   b) affects eBay as well.
   c) has extremely increased on eBay sites in the last few years.
   d) has led to a loss of trust in eBay.

b **Vocabulary**

1. Explain the following words or phrases in English.
   a) eBay (headline)
   b) dot.com companies (line 14)
   c) economist (line 32)
   d) cyber crime (line 52)
   e) hackers (line 59)

2. Find the corresponding adjective.
   a) success (line 3)
   b) created (line 18)
   c) demand (line 19)
   d) differences (line 66)

3. Find synonyms for these words.
   a) rising (line 4)
   b) annual (line 11)
   c) broken (43)
   d) secure (line 51)
   e) problems (line 56)

R **c** **Reading comprehension** → SF 1

Answer the following questions in your own words as far as possible.

1. How does eBay's business concept work?
2. How did eBay start?
3. What facts mentioned in the text show that eBay is very successful?
4. What is the special idea behind eBay?
5. What are the dangers of internet shopping?

d **Summary** → SF 6

Using your own words, give a summary of the text. It should be about 250 words long.

M **e** **Translation** → SF 10

Translate lines 7 to 30. (By now … selling goods on eBay.)

P **f** **Composition** → SF 5

Comment on the following statement, using your own knowledge and ideas.

"In the future we will do all our shopping from home so all supermarkets and stores will have to close down."

1 Listening comprehension: Online shopping

R **a** Listen and find out if the following statements are true or false. Correct the false statements.

1. The article is about selling household appliances online.
2. Most work in the warehouse is done by remote-controlled machines.
3. Ocado delivers to customers throughout England.
4. Ocado's turnover has increased constantly.
5. Ocado makes good profits.
6. In the future Ocado plans to sell goods worth £ 3bn.
7. A great advantage is that deliveries are made around the clock.
8. The average order is £ 75.
9. Experts are sceptical because most companies in that business have not been successful.
10. It was easy for Ocado to find investors.
11. Ocado needs to invest further £ 170m to make profits.

P **b** Copy the following mind map into your exercise book. → **SF 9**
Then complete the mind map, using your own knowledge and ideas.

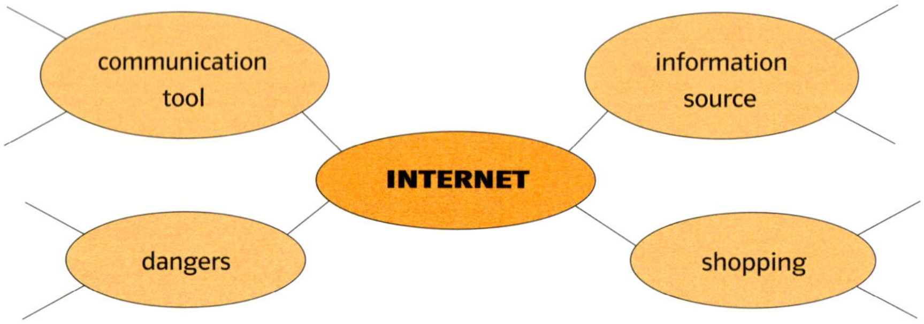

c With the help of your mind map prepare and give an oral presentation on the Internet. → **SF 13**

2 Presentation: Online shopping versus mall shopping

Online shopping is becoming more and more successful and not only young people do their shopping online nowadays. However, does it really offer that many advantages or are there disadvantages, too?

Work with a partner or in small groups and compare online shopping to mall shopping. Find advantages and disadvantages for both forms and prepare a poster. Present it to your class. → **SF 22**

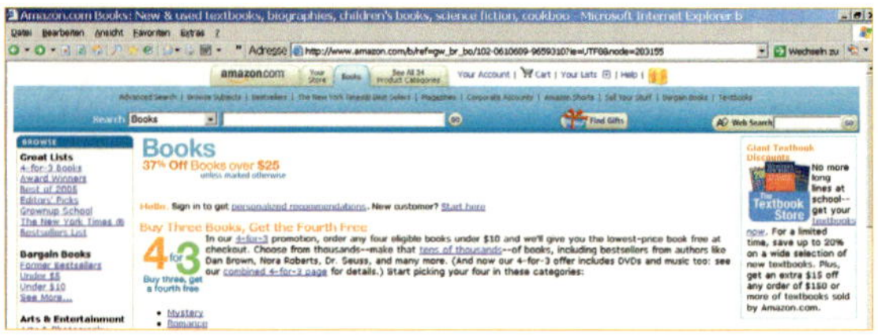

1 Describing a diagram: US online retail sales → SF 21

P ■ Describe and analyse the following diagram showing U.S. online retail sales in billion US-Dollars.

U.S. online retail sales in billion US-Dollars

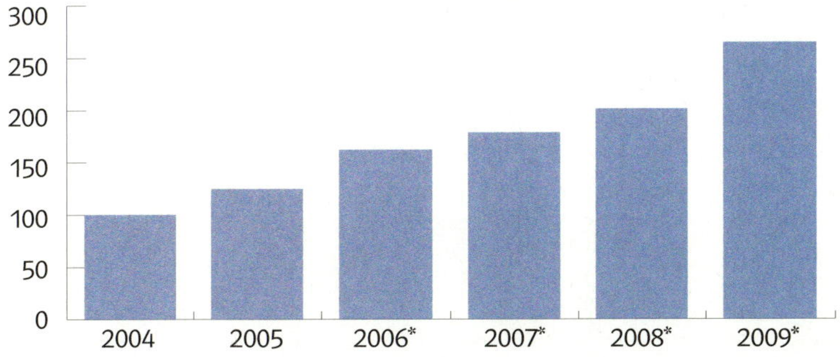

* forecast

2 Describing a cartoon: Online shopping → SF 20

P ■ Describe what you can see in the following cartoon and try to explain its message.

Fame, fun, fortune

(A) If you look in a newspaper in London or New York, there is a good chance you will see the smiling face of Richard Branson, airline owner, record salesman and extraordinary businessman. According to the latest *Sunday Times* "Rich List", Richard Branson is estimated to be worth £2,600 million which makes him the sixth wealthiest person in Britain.

(B) Richard Branson started business in the seventies with a small mail order record firm. Since then he has gone into many other fields of business and he is now the owner of the Virgin Group, an international company with a turnover of £4.5 billion. More than 200 companies employ over 25,000 people.

(C) And his British-based company is still expanding. After he had founded Virgin Atlantic Airlines in 1985, which operates all over the world, Richard Branson launched a further airline called Virgin Express which offers flights to European cities at low prices. Moreover, Virgin's business on the ground is expanding, too. Nowadays, for example, Branson's chain of shops offer a massive selection of music, video and computer games not only to customers in Europe but also in the United States and Asia. In addition, the Virgin Group has invested in hotel accommodation in exclusive and luxurious locations, including Branson's own Caribbean island. In this way, the Virgin Group has become a mini-empire consisting of hundreds of different companies and a number of international joint ventures in which Virgin keeps management control.

(D) According to Richard Branson the reason for the company's success lies in the fact that Virgin has always done things differently. Moreover, he has always tried to keep the small business atmosphere. A typical example of this can be seen in the way he operates his airline. Richard Branson admits that a plane is a plane, and at the end of the day most of his competitors will be able to match any hardware that Virgin offers on the market. However, Richard Branson claims that software – human service – is what is most important, and the key to success is to have a highly motivated staff. To achieve this, Virgin has divided its cabin crew members into teams, who then feel directly responsible for their work and their company. As a result, Virgin Atlantic Airlines was the first airline to offer first-class service at business-class prices. Richard Branson has introduced a similar strategy in his other business branches. He motivates the individual managers by making them feel that they are the real bosses and then lets them run the business as they like. When a flight attendant came to him and suggested a wedding business, Richard Branson told her to go ahead with the idea. He even put on a wedding dress to give the new company publicity.

(E) In addition to his own business activities, Richard is a trustee of several charities including the Virgin Healthcare Foundation, a leading healthcare charity which was responsible for the launch of a health education campaign relating to AIDS in 1987. The Foundation has also become involved in a campaign called Parents

Against Tobacco, which aims to restrict tobacco advertising and sponsorship in sport. His help in the initial funding of charity projects helped that organisation to raise over £100 million.

(F) Richard's private life is also very unusual: his record breaking crossing of the Atlantic in a speedboat and his most recent record crossing of the English Channel in a £150,000 amphibious sports car in 2004 are just two of his highly publicized adventures.

(G) Richard Branson still loves what he does and he can't see himself slowing down just yet. He says he has a fantastic team of people who do a tough job and seem to enjoy running the businesses with him. They all try their best to keep up with any changes and trends, and by doing so find themselves entering new areas of business all the time. Keeping with the pace and continuing to experiment into new ventures means that he is doing different things all the time anyway. This all seems to work and the Virgin Group has continued to grow successfully. Richard Branson constantly comes up with new ideas: a reality show based on the "Big Brother" idea, the discount airline Virgin USA and ultimately space travel by "Virgin Galactic" at affordable prices. But the next phase of growth for the Virgin brand will be starting businesses in new countries and markets, rather than expanding the range of things that they do. By worldwide standards, Virgin is still a small group of companies, but the brand is a major multinational one. That gives them lots of opportunities to create a truly global presence, and to do some good, too.

(H) Some years ago, Richard Branson was awarded a knighthood in the Queen's Millenium New Year's Honours List. Well done, Sir!

790 words

R a Reading comprehension → SF 1

1. The heading of the text is "Fame – fun – fortune". Give an example of each item from the text.
2. With reference to the text say what the photo shows.
3. Find three examples of different fields of business in the text.

R b Multiple matching

Match the following headings with the suitable paragraph (A)–(H). There are two headings that do not fit.

1. Charity activities
2. The future development
3. Virgin's business still developing
4. A famous and rich businessman
5. The Virgin Group in figures
6. From airlines to wedding businesses
7. Private adventures
8. Sir Richard
9. The reasons for his success
10. Managers as bosses

M c Mediation → SF 10

Übertragen Sie die Zeilen 24 – 42 ins Deutsche. Übersetzen Sie nicht Wort für Wort und schreiben Sie ganze Sätze.

P d Composition → SF 5

Comment on the following statement using your own knowledge and ideas.
"For employees there are more advantages and benefits in working for a small company than working for a large international concern."

1 Listening comprehension: The history of Airbus

R ■ Listen and find out if the following statements are true or false. Correct the false statements.

1. Airbus was founded by a consortium of French, German and Spanish companies.
2. The first Airbus had four engines.
3. British Aerospace joined in the late 70's.
4. Airbus had to make its production more efficient because there were a lot of orders.
5. In 2005 Airbus managed to get more orders than its main rival.
6. Centres of Excellence were established to stay competitive.
7. Airbus has already sold 555 A380.

2 Composition: Airbus statistics

P ■ Comment on the following mind map. → SF 5

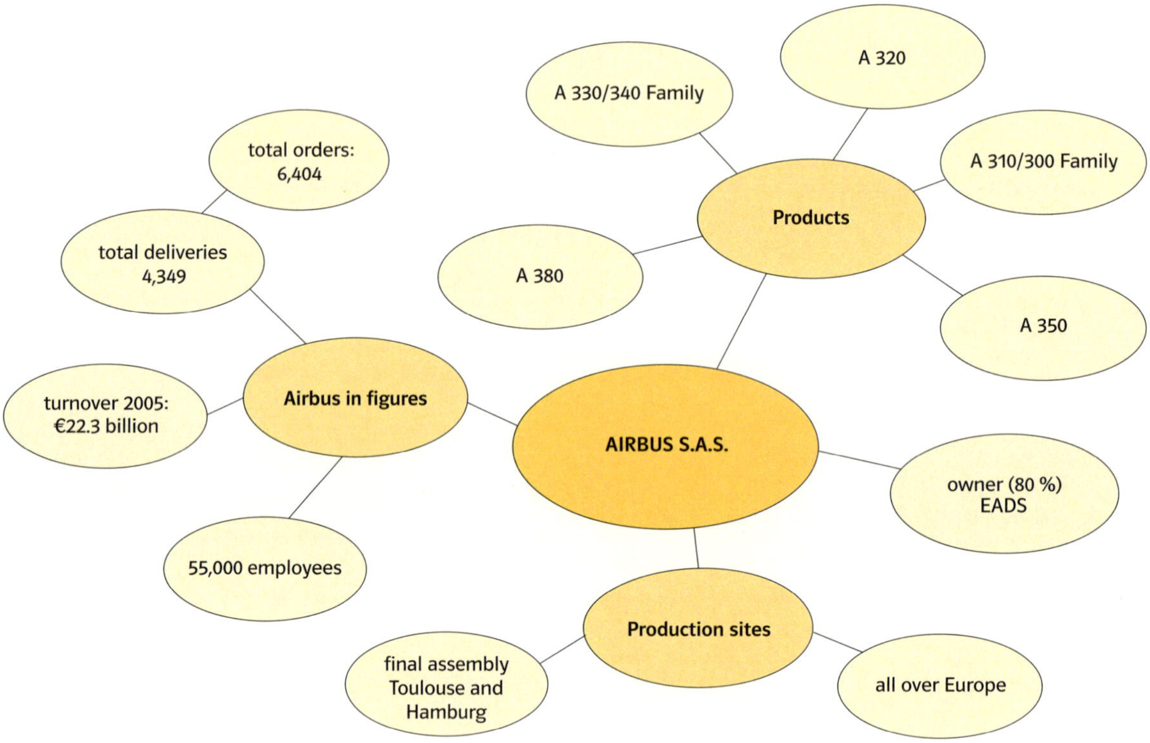

P **a** Describe the following advertisement in about 80 words. → **SF 20**

I **b** Role play: → **SF 13+14**

Situation: Four colleagues are working on new ideas for a drink that your international food company wants to introduce on the market. Each colleague has a different drink in mind.

Your choice of drinks: A soft drink, a milk shake, a low-calorie drink and a low-alcohol drink.

Presentation: Introduce yourself, present the drink that you have developed and try to convince the others that your business idea is the best.

Discussion: Discuss the pros and cons of each drink and try to reach an agreement on which drink should be introduced.

Unit

20 Globalisation

Globalisation

(A) Globalisation is a controversial issue. It has generated large protests around the world, by people who feel that it benefits only the rich. Yet there are others who claim that it offers real solutions to global poverty. There are arguments for both sides. In practice, globalisation has the potential to do both good and harm.

(B) The World Bank defines it as 'the growing integration of economies and societies around the world'. It sounds simple, but the processes of globalisation involve changes to many different aspects of society: from communications, to travel, to economics, to government.

(C) Globalisation represents a particular economic theory based on the belief that a liberalised, free market is desirable. Supporters of a free market economy believe that this approach promotes healthy competition. Countries can specialise in producing the goods they make most efficiently and export these goods to other countries without restrictions. Only those who produce goods efficiently and at a competitive price will survive.

(D) Globalisation impacts virtually every aspect of life including religion, culture, diet and family life. One of the most striking impacts of globalisation is the increasing connection of economies and cultures, making the world seem smaller. A decision made in Japan can employ thousands in another country, or render them jobless. A flu virus in Asia can affect tourism and business worldwide.

(E) Economic and technological change are the major forces driving globalisation. Worldwide, trade is increasing as global markets become more closely linked through improved communication technologies like the internet. These same technologies are bringing distant communities together, and making it easier for corporations to move their operations to areas where costs are low. Cheaper air travel because of technological improvements means more people are travelling for business or pleasure, making workers more mobile.

(F) The integration of global economies provides enormous potential for all economies to expand into different parts of the world – to both sell and buy new products and services. While it can be difficult to enter new global markets, globalisation is giving poor countries some chance of gaining the benefits of world trade. If poor countries could get access to just one percent more of the value of world trade, it would lift well over 100 million people out of poverty through the economic growth and jobs created. As multi-national corporations move production to benefit from cheaper labour costs in poorer countries, they are creating jobs for people who previously had few opportunities.

(G) Technological advancements mean quick and affordable worldwide travel is now available, particularly to people from wealthier nations. This is helping to boost tourism in many developing countries. The internet and other improved telecommunication facilities enable information to be sent from one corner of the planet to another in seconds. This has brought together people from vastly different backgrounds and cultures. Ideas are shared, business is conducted internationally and those affected by a disaster can call for, and get, help faster than ever before.

(H) One concern is the effects of constant pressure from large corporations to lower prices and shorten production times for goods. Although jobs are created, poor labour standards and cost cutting processes in developing countries mean wages can be low, conditions are often dangerous and workers' rights are violated.

(I) Critics of free-market economics note that industries in developed countries did not develop in these open conditions. Instead, they were protected locally (often with subsidies) and their markets were closed to outside competition until new industries became established. Yet poor countries are now expected to perform and survive in new global markets without first starting from an established base and without adequate support or concern for local interests. This makes it difficult for developing nations to benefit from the increased trade opportunities offered by globalisation.

(J) Rich countries primarily produce manufactured goods for the world market, which hold their value and command good prices. Developing countries generally produce commodities – raw materials like cotton, coffee, cocoa, metals and minerals – cheaply and efficiently, but do not have the current capacity to produce manufactured products. Unlike manufactured goods, commodities are extremely vulnerable to wide price swings whenever there is a shortage or oversupply. Countries which depend on only a few vital commodities for income may struggle if the global price suddenly plummets.

712 words

(worldvision.com.au)

R a Multiple Matching

Match the following headings with the suitable paragraph (A)–(J). There are two headings that do not fit.

1. Poor work conditions
2. More travel and faster communication
3. Is globalisation good or bad?
4. The free market economy
5. What is globalisation?
6. The internet
7. No protection to get started
8. The World Bank
9. Vulnerable to world prices
10. More jobs
11. What's driving it?
12. The impact on our lives

b Vocabulary

1. Explain the following words or phrases in English.
   1. free market (line 17)
   2. competitive price (lines 22/23)
   3. multi-national corporations (lines 55/56)
   4. subsidies (line 82)
2. Find the corresponding adjective.
   1. societies (line 9)
   2. represent (line 14)
   3. life (line 25)
   4. access (line 52)
3. Find the corresponding verbs.
   1. solutions (line 5)
   2. competition (line 18)
   3. advancements (line 60)
   4. internationally (line 70)

R c Reading comprehension → SF 1

Answer the following questions in your own words as far as possible.

1. Why is globalisation a controversial issue?
2. Which advantages does a free market economy bring?
3. Which factors enable globalisation?
4. How can developing countries benefit from globalisation?
5. How do some large corporations exploit workers in developing countries?
6. Why is it dangerous for countries to rely on commodities?

M d Mediation → SF 10

Übertragen Sie den Inhalt der Abschnitte F und G ins Deutsche. Übersetzen Sie nicht Wort für Wort und schreiben Sie ganze Sätze.

1 Listening comprehension: Cheap labour

R ■ Listen and find out if the following statements are true or not true.
Correct the false statements.

1. Chenda works as an office clerk at a clothing factory.
2. Chenda's wage is very low and therefore illegal.
3. Most interviewed workers weren't satisfied with their working conditions.
4. The workers are happy that there are at least enough fans in the factory.
5. The work in the washing room is unhealthy because workers aren't protected well enough.
6. The factory installed new, safer washing machines last year.
7. 200 people have been fired since 2004 because they demanded better wages.
8. The workers plan to strike for better working conditions.
9. Fortune admits that they fired workers because of union activity.
10. According to the International Labour Organisation (ILO) working conditions at Fortune are good.

2 Describing a cartoon: The rich and the poor

P ■ Describe what you can see in the following cartoon and try to explain its message. → SF 20

1 Describing a diagram: The world's poorest and richest nations

P ■ Describe and analyse the following diagram showing income per person (GDP (nominal) per capita) in the world's 20 poorest and 20 richest nations. → SF 21

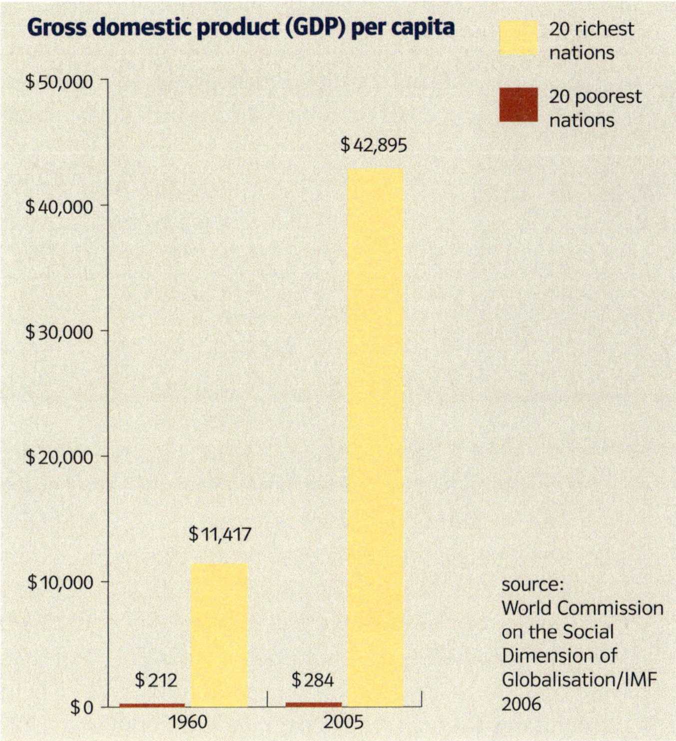

2 Composition: Globalisation

P ■ Comment on the following mindmap using your own knowledge and ideas. → SF 5

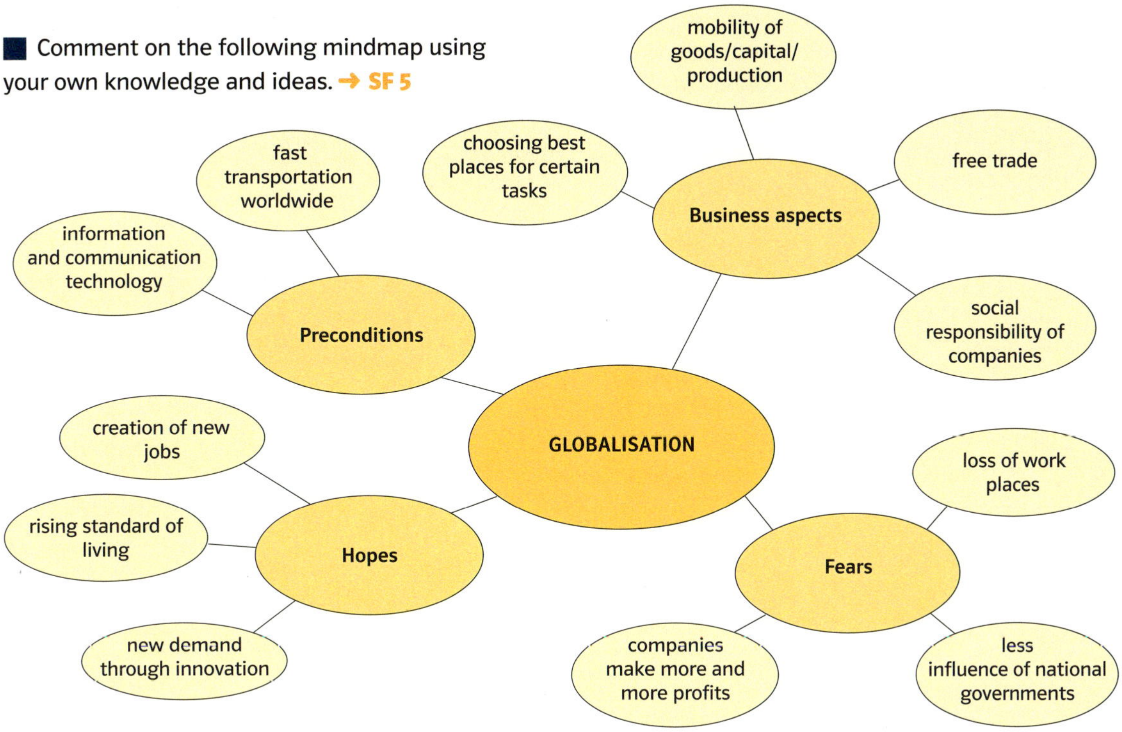

Unit

21 Genetic engineering

UK breakthrough as human embryo cloned

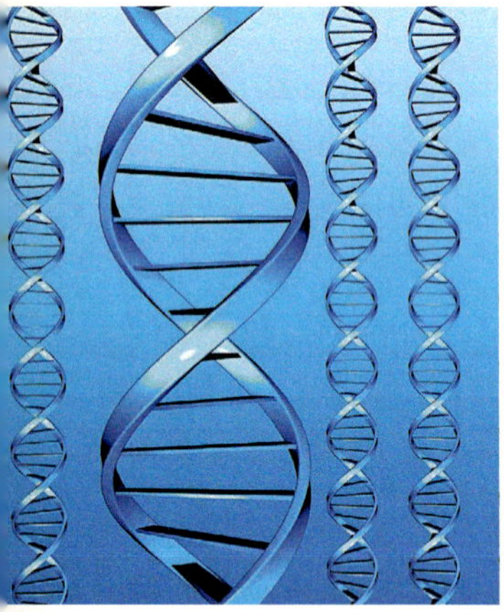

(A) Scientists in Newcastle have successfully cloned a human embryo, a breakthrough that places Britain at the forefront of the highly controversial field of embryonic stem cell technology.

(B) The clone was created as part of the Newcastle group's research into new treatments for diabetes. The team, lead by Miodrag Stojkovic at Newcastle University and Alison Murdoch at the Newcastle NHS Fertility Centre, was the first in Europe to be given the go-ahead to clone embryos for research last year, after being granted a licence by the Human Fertilisation and Embryology Authority (HFEA).

(C) The Newcastle team's work was praised by scientists who believe embryonic stem cells, which can be extracted from cloned embryos, will pave the way to cure some of the most intractable medical conditions, including Parkinson's, Alzheimer's disease and even paralysis. But the development has reignited the debate over the ethics of creating early-stage embryos for research, with critics accusing the scientists of trivialising life.

(D) Britain is one of a handful of countries to permit the cloning of human embryos for research, or therapeutic cloning. Several countries throughout Europe have introduced bans or severe restrictions on cloning research and the UN is pushing for a global ban.

(E) To create the clone, the team collected 36 surplus eggs donated from 11 women undergoing IVF (in-vitro fertilisation) treatment. Each egg had its nucleus replaced by a whole human embryonic stem cell from the UK stem cell bank. The eggs were then given a brief electrical shock to kickstart the growth process. From 10 of the eggs, the researchers were able to create three very early stage embryos, with one developing into a blastocyst, a ball of cells no bigger than the head of a pin. Attempts to extract stem cells from the blastocyst failed, however, because the clone did not survive beyond five days. The experiment was done as a proof of principle, to see if eggs collected from women undergoing IVF treatment would be healthy enough to produce clones.

(F) The team has gained permission from the HFEA to create more clones from cells taken from patients with Type 1 diabetes. The stem cells from those clones will carry the genetic defect that causes diabetes and will enable scientists to study the disease in far greater detail than ever before. "With these stem cells we will be able to study the very roots of the disease," said Dr Stojkovic.

(G) The group has also been granted permission from an ethics committee to seek eggs from women who produce too many after IVF treatment. During the experiment the team found that only fresh eggs used within an hour after being collected were good enough to produce clones. "We know from our analysis of IVF data that if a woman has 20 eggs, her chance of getting pregnant is no higher than if she has 18, so on that basis we can be very confident that the women who donate would not significantly decrease their chances of having a baby," said Prof Murdoch. Dr Stojkovic emphasised it was a technique to get stem cells. "We're not doing this to clone human beings," he said.

(H) Pro-life groups criticised the work as unethical. "This work trivialises life. An embryo, no matter how small, no matter how it is created, is a human life. We must respect that," said Josephine Quintavalle of Comment on Reproductive Ethics (CORE). The HFEA's right to grant permission for the work is being tested in the high court after lawyers for Christian Fellowship won a judicial review of its decision.

IVF = in-vitro fertilisation (künstliche Befruchtung)

603 words

(The Guardian, 17 May 2005, abridged)

R **a** **Multiple choice**

Find the most suitable option

1. The text is about
   a) cloning human beings.
   b) using embryonic stem cells to cure illnesses.
   c) cloning animals.
   d) new treatments for Parkinson's and Alzheimer's disease.
2. British scientists
   a) are leading in the field of stem cell technology.
   b) have found a way to clone human beings.
   c) had to stop their stem cell research.
   d) have found a way to cure diabetes.
3. Who allowed to clone human embryos for research?
   a) The British government
   b) The Human Fertilisation and Embryology Authority
   c) The Christian Fellowship
   d) The UK stem cell bank
4. Most countries in Europe
   a) are working in the field of embryonic stem cell technology as well.
   b) are fighting for a global ban of cloning research.
   c) do not allow cloning research.
   d) are trying to make therapeutic cloning successful.
5. The team of Dr Stojkovic and Prof Murdoch
   a) is now using cloning to learn more about diabetes.
   b) is now trying to find a cure for Parkinson's and Alzheimer's disease.
   c) is now trying to improve IVF treatment.
   d) is now trying to extract stem cells without genetic defects from diabetes patients.
6. The group Comment on Reproductive Ethics (CORE)
   a) is fighting cloning in court.
   b) allows stem cell research only if no human beings are cloned.
   c) puts above all absolute respect for the human embryo.
   d) organised demonstrations against stem cell research.

R **b** **Reading comprehension** → SF 1

1. How could we benefit from embryonic stem cell technology?
2. How were the reactions to the success of the Newcastle team?
3. Why does Prof Murdoch believe that women who donate eggs will not be less likely to have a baby?
4. What does Josephine Quintaville think about the Newcastle team's work?

M **c** **Mediation** → SF 10

Übertragen Sie den Inhalt der Abschnitte F und G ins Deutsche. Übersetzen Sie nicht Wort für Wort und schreiben Sie ganze Sätze.

P **d** **Describing a cartoon**

Describe what you can see in the cartoon on the right and try to explain its message. → SF 20

"Psst – rumour has it that the boss is a self made man."

22 Peacekeepers

UN blue helmets earn applause – and censure

(A) When United Nations peacekeepers opened fire on alleged gang members in a slum of the Haitian capital some six months ago, they also killed or injured dozens of innocent men, women and children. The incident was widely condemned by human rights activists who accused the troops of being cold-blooded murderers.

(B) By contrast, thousands of miles away in Liberia, U.N. peacekeepers have become unsung heroes who not only made it impossible for despot Charles Taylor to return to power, but also helped Liberians elect their first female president by putting an end to years of civil strife and anarchy.

(C) While many would perceive these two episodes as representing a mixed record of performance by the world body's peacekeeping missions in 2005, U.N. peacekeeping officials disagree. "It has been a great year," says David Harland, a senior official at the U.N. Department of Peacekeeping Operations (DPKO), describing the end of the U.N. mission in East Timor as "a great achievement" and the departure of U.N. troops from Sierra Leone as a "landmark event".

(D) After helping the voluntary repatriation of more than 200,000 people who fled to West Timor to escape the violence after the 1999 referendum on independence for East Timor, the U.N. is scheduled to finish six years of humanitarian operations on the island on Saturday.

(E) Similarly, the U.N. is winding up its mission in Sierra Leone after six years of peacekeeping there to put an end of years of vicious civil war. The West African nation is now poised to hold democratic elections in 2007.

(F) But what about the charges of sexual abuse of women and girls in the Democratic Republic of Congo and other countries, and the killing of innocent civilians in Haiti by the peacekeepers? Officials say so far the U.N. has completed more than 200 investigations into charges of sexual abuse and that it has arranged stricter training for peacekeepers on their conduct. In addition, the soldiers accused of sexual abuse are facing prosecution in their home countries.

(G) Currently, U.N. peacekeeping troops are deployed in 17 countries around the world, with more than 80,000 personnel, including troops, police and civilian staff. The number of operations has grown rapidly. During the first five years of the 1990s, there were more U.N. peacekeeping operations than in the previous 45 years.

(H) Officials admit that the July incident involving U.N. troops in Haiti might have led to civilian casualties, but still defend their actions. "Haiti is problematic and a failed state," says Harland. "It's out of control." Like the interim Haitian government that replaced the popularly elected president Jean-Bertrand Aristide, the U.N. forces in Haiti have branded some opposition activists as "gangsters", and have tried to deal with them militarily. U.N. officials justify military action against rebels by arguing that the mission in Haiti had to abide by the Security Council's verdict, which considers the interim government as legitimate. "The government is our partner in running the elections in Haiti. Members of the Security Council support that government," Harland explained. "They are cooperating with us."

(I) On a question about the role of peacekeepers as being neutral, he said neutrality was something that was not central to peacekeeping operations, adding that the mission had to act in accordance with the Security Council's decisions. However, Harland thinks that it's "awful that innocents are harmed".

(J) Independent observers link the killings in Haiti and sexual abuse in other countries to lack of proper training for the peacekeepers. "About 80 percent of the peacekeeping troops come from the poor countries because rich countries are unwilling to contribute any of their own," says Jim Paul, director of the Global Policy Forum, an independent group that closely observes U.N.-related developments.

622 words

(United Nations, Dec 30, 2005)

R **a** **Multiple Matching**

Match the following headings with the suitable paragraph (A)–(J). There are two headings that do not fit.

1. Reasons for misbehaviour
2. Haiti and its problems
3. Peace for Sierra Leone
4. Being impartial?
5. The failure
6. Back to democracy
7. Origin of UN troops
8. Sexual misconduct
9. A successful year
10. Getting West Timor back on track
11. Civilian casualties
12. UN Peacekeeping around the world

R **b** **Multiple choice**

Find the most suitable option.

1. David Harland describes 2005 as a great year
   a) because there was an election in Sierra Leone.
   b) because UN Peacekeepers were celebrated as heroes.
   c) because UN Peacekeeping had a mixed record.
   d) because UN peacekeeping was successful in several countries.
2. 'Voluntary repatriation' in paragraph D means that
   a) people will be allowed to go back home next year.
   b) people were allowed to go back to their home country.
   c) refugees will be allowed to return because there is peace.
   d) refugees had to return back home.
3. People in Sierra Leone
   a) are ready for democratic elections in 2007.
   b) are afraid that there could be no elections in 2007.
   c) accused some UN soldiers of sexual misconduct.
   d) are waiting for an end of the civil war.
4. How has the UN reacted to charges of sexual abuse?
   a) 200 soldiers have been dismissed.
   b) The soldiers have been sent to prison.
   c) Training for peacekeepers has been improved.
   d) The soldiers have been prosecuted by the UN.

M **c** **Mediation** → SF 10

Übersetzen Sie bei der Beantwortung folgender Fragen nicht Wort für Wort und schreiben Sie ganze Sätze.

1. Wieso starben in Haiti unschuldige Menschen?
2. Wieso war die UN Mission in Liberia ein Erfolg?
3. Wie erklären unabhängige Beobachter die Vorfälle in Haiti?

P **d** **Composition** → SF 15

Comment on the following statement, using your own knowledge and ideas.
"The UN only works for the interests of developed countries."

Healthy eating

Fast food

(A) Teachers have called for a ban on junk food advertising aimed at children and warned that they could not reverse young people's unhealthy eating habits alone, as the government put schools at the heart of its plans to tackle childhood obesity.

(B) The call comes as the government today launched a four-step blueprint for improving pupils' food consumption and education about healthy eating.

(C) It pledged to introduce a "whole school" approach to reforming what school pupils eat and how they exercise. Schools will be encouraged and supported to teach healthy lifestyles across the curriculum, providing healthy options in school canteens, expanding the opportunities for exercise and introducing more options to bring health workers into schools permanently.

(D) The general secretary of the National Union of Teachers, Steve Sinnott, said: "I welcome the government's healthy living blueprint for schools. It is a good first step in tackling the pernicious influence of the fast food manufacturers' dominance over young children's lives.

(E) I call on the government to ban junk food advertising aimed at young children. The government needs to understand the overwhelming influence of rampant commercialisation. Obesity and poor nutrition affects learning."

(F) John Dunford, general secretary of the Secondary Heads Association, said headteachers would welcome the government's scheme, but added: "We're not just trying to change schools' activities; we're changing national habits. A very important part of that is a more responsible food industry.

(G) While schools are willing to play their part in encouraging healthy living by young people, food companies continue to produce unhealthy foods and target children with advertising. The government must put more pressure on the food industry to conform to higher standards of nutrition."

(H) He added: "There is a lot more choice now in schools. But children choose the less healthy options ... and the shop around the corner sells the unhealthy foods that many children want."

(I) Chris Keates, acting general secretary of the National Association of Schoolmasters Union of Women Teachers, the largest union representing teachers throughout the UK, also welcomed the drive, but added: "Schools can and should lead by example, but it is essential not to fall into the trap of raising expectations that schools can compensate entirely for the dietary and lifestyle deficiencies which are sadly a part of life outside school."

(J) Teachers' concerns reflect the growing trend of pupils opting out of school dinners in favour of buying lunch at the local take-away or corner shop.

(K) Parents and teachers have previously expressed concerns about the marketing tactics of some food companies. In July last year a row broke out after it emerged that schools were giving away McDonalds vouchers as rewards for good attendance. The debate followed a similar argument in March when Cadbury's announced a multi-million pound campaign offering new school sports equipment in exchange for chocolate wrappers - the campaign was later dropped.

(L) Today's blueprint offers a four-pronged attack – changing the food on offer, encouraging children to learn about what they are eating, and how it affects their body and introducing more exercise. The government will also update its healthy school dinners' guidelines - first set in 2001, but now in need of revision to take into account the latest research on salt consumption and saturated fats.

(M) The minimum nutritional standards for school lunches, the first for 20 years, are now met by 83 % of secondary schools. Eight out of 10 schools now provide free drinking water at the beginning of the day and 91% of schools provide vegetables and/or fruit on most days.

(N) The education secretary, Charles Clarke, said: "Good health and effective learning go hand-in-hand, a healthy body leads to a healthy mind. The partnership of parents and school is critical in shaping good health in children, and schools are well-placed to lead by example."

625 words

(The Guardian, 6 September 2004)

R a Multiple choice

Find the most suitable option.

1. Paragraph C: Instead of simply saying "providing healthy options" you could also write:
   a) by providing healthy options
   b) when providing healthy options
   c) despite providing healthy options
   d) as well as providing healthy options
2. According to Steve Sinnott the government
   a) should talk to the producers of fast food
   b) is on the right way, but has to do more.
   c) has understood how fast food affects learning.
   d) dominates young children's lives
3. According to paragraphs E and N teachers are against the consumption of too much fast food because children …
   a) become fat
   b) are influenced by fast food advertising
   c) do not learn as well as they could
   d) do not get enough exercise

R b Reading comprehension → SF 1

Answer the following questions in your own words as far as possible.

1. How does the government try to change unhealthy eating habits of school children?
2. What does Steve Sinnott think about the blueprint?
3. What does the food industry do to prevent children from becoming obese?
4. Why do more and more pupils buy their lunch at corner shops outside school?

M c Mediation → SF 10

Übertragen Sie die Abschnitte A – D ins Deutsche. Übersetzen Sie nicht Wort für Wort und schreiben Sie ganze Sätze.

P d Describing a cartoon → SF 20

Describe what you can see in the cartoon and try to explain its message.

Well done Son. You've put on another inch since last week.

Multiculturalism

How British do you want to be?

Trevor Phillips, chairman of the Commission for Racial Equality, has called for an end to multiculturalism in Britain. Separateness is over, he said last week. It is time to rediscover the nation's belief in 'core British values' and integrate under an umbrella marked with the colours of the Union flag.

The Observer asked people from across the cultural and racial spectrum, all British passport holders, for their point of view.

(A) Sir Gulam Noon – Businessman

I hate the words 'ethnic minority'. Those who are born in this country from Asian communities feel they are British, and they are British. They have been educated in British schools and universities; the way they walk, talk and dress is British.

The host community doesn't object to you keeping your culture intact. I remember that in the Fifties and Sixties there were only a handful of mosques and now there are more than 1,000. Temples, too, have sprung up and the Neasden temple welcomes a large number of English visitors.

The second generation Asian immigrants are British and by the same token they are Muslims and Hindus as well. They have a dual identity. I consider myself British. But my religion is Islam and I am proud of my heritage. My Britishness doesn't mean I love India any less. For goodness sake, be loyal to the country in which you live.

(B) Parmjit Dhanda - Labour MP for Gloucester

My parents' generation were not presented with a choice about living in a multicultural society or an integrated one.

We may celebrate people like my mum for building the NHS, but the sad truth is that when she and many other Asian women like her arrived here in the 1960s they had few options other than to clean hospital toilets for a living. They felt marginalised and were ushered into silos to live – greater integration was never on offer for those here to 'nick our jobs'.

Whether it was Southall, as in my parents' case, or Brixton, Handsworth or parts of Oldham, they found safety in numbers. Since then, little by little, around the country we have seen diversity work.

There is nothing unique about my experience. I grew up in West London, support Liverpool at football, India at cricket and proudly represent Gloucester in parliament. Confused? No, I'd say representative of the British experience in the here and now.

(C) Charles Bailey – Record producer

You can't change history and, like it or not, colonialism and immigration have brought a lot of people from different cultures into the 'British' mix. So multiculturalism is not the issue. What we should be talking about is integration. Ignorance is the main problem and we could get rid of that just by getting to know each other better. Ethnic minorities need to get more involved in mainstream society so they don't seem so mysterious to the home Brits.

Black people have managed to break down a lot of barriers over the years and maybe some of the Asian and Muslim communities who are coming under a bit of pressure can learn from our experience. Take football, for example. Players like Ian Wright have done an enormous amount to get racism off the terraces, because it's very hard to hate someone if they are your star player. You have to get involved, but that doesn't mean you have to stop being who you are.

(D) Fiona MacTaggart – Minister for race equality and community policy

It is often the places that are truly multicultural that are most creative. The differences between cultural practices and attitudes spur imagination and inform reflection and understanding.

Multiculturalism can easily be no more than a careless acceptance of people living side by side but leading wholly separate lives which never interact. But actually it has been their interactions and adaptations which have contributed so much to Britain today.

Citizens should enjoy, celebrate and understand the culture that they have inherited and use it to contribute to British society today – only dead cultures cease to evolve.

The debate now must be about how positive integration can ensure no one is left behind, and how we can realise the strength from our diversity as a society.

(E) Christine Yau – Chair, Soho's Chinatown community centre

If you leave your own country you have to be prepared to integrate. Every Chinese person has the intention to go mainstream. For some of the older generation it can be difficult to do this because of language barriers. But it's the main objective to get their children to go mainstream. A lot of Chinese kids are guided by their parents to follow traditional British education. At the same time they are sent to Saturday and Sunday school to learn about the Chinese culture and in the home they speak their own Chinese dialect. This method of balancing the two cultures has worked for us. There aren't many Chinese figureheads out there because it's in their nature to be conservative. They don't court publicity. They are looking to get on with their lives. But I see more youngsters getting involved in community work.

830 words

(The Observer, 11 April 2004, abridged)

R a Multiple Matching

1. Match the photos with the appropriate name and text.
2. Match the person with the correct statement according to the text.

a) Sir Gulam Noon b) Parmjit Dhanda c) Charles Bailey d) Fiona MacTaggert e) Christine Yau

1. The culture of an ethnic minority should act as the basis for creating a more interesting British society.
2. Strong separation of the different cultures is the best way to keep peace.
3. The first generation of immigrants only had few job and housing alternatives.
4. Black sports stars have already broken down a lot of prejudices which British people may have had towards ethnic minorities.
5. The number of non-Christian places of worship in Britain has increased dramatically over the last sixty years.
6. Immigrants should give up their original identity and become British.
7. In some cases children from ethnic minorities are strongly encouraged to follow a typically British type of education.

R b Reading comprehension → SF 1

Answer the following questions in your own words as far as possible.

1. What does Trevor Phillips mean when he calls for an end of multiculturalism?
2. What does Sir Gulam Noon mean when he uses the term "dual identity" (line 27)?
3. Which experiences did Parmjit's mother make when she came to Great Britain?
4. How could stars, according to Charles Bailey, reduce racism?

M c Mediation → SF 10

Übertragen Sie den Abschnitt E ins Deutsche. Übersetzen Sie nicht Wort für Wort und schreiben Sie ganze Sätze.

P d Composition → SF 5

A reporter from The Observer asks you to state your opinion about multiculturalism in your country. Write a short text with regard to the statements you have read in the newspaper article.

Appendix

Overview

Skills files

Overview

File 1 Reading Comprehension

1. In order to have a general idea of the text, first look at the heading, pictures or diagrams. If necessary, read the first paragraph(s).
2. Try to remember what you already know about the topic. This will help you to understand the text.
3. Read through the text quickly in order the get a general impression of its contents (skimming, see Skills Files 2). Check if your first impression of the text was appropriate.
4. Read your tasks carefully. Then read the text (or part of it) again and look for the pieces of information that are required (scanning, see Skills Files 2). If possible, mark important details in the text or note them in the margin. Otherwise make notes on a separate piece of paper.
5. Try to find the meaning of important unknown words from the context or with the help of your dictionary. → SF 3+4 Note definitions or explanations if necessary.
6. Use your notes to do the tasks.

File 2 Skimming / Scanning

■ Skimming and scanning are both fast reading techniques which will help you to read texts more quickly and effectively.

■ What is skimming and how does it work?

Skimming is used to get a general impression of the main ideas of a text. It will help you to decide if the text is interesting and whether you should read it in more detail. You often use this technique when you read a newspaper. Skimming is done at a speed three or four times faster than normal reading.

1. Read the title, subtitles and subheading to find out what the text is about.
2. Look at the pictures or images.
3. Read the first and last sentence of each paragraph.
4. Don't read word-by-word and let your eyes run quickly over the text.
5. Think about the meaning of the text.

■ What is scanning and how does it work?

The purpose of scanning is to find specific information (e.g. key words or ideas) without reading through the whole text. We often use this technique when looking up a word in the telephone book or dictionary.

1. If you need to do some tasks (e.g. answer questions) read them carefully first.
2. Let your eyes move quickly down the page until you find what you are looking for. Don't read word-by-word.
3. Mark important details in the text or note them in the margin. Otherwise make notes on a separate piece of paper.
4. Try to find the meaning of important unknown words from the context or with the help of your dictionary.
5. Use your notes to do the tasks.

File 3 Surviving without a dictionary

The following techniques will help you to 'survive' without a dictionary, when you come across words you don't know:

- Always read a text once through without looking up any words. Try to understand the gist, i.e. the basic meaning of a text. You don't need to understand all the details at first and sometimes the gist alone is enough.
- See whether the words and sentences before and after the word (the context) can help you to understand it! Can you guess the meaning by the context? Or can you understand the sentence without needing to understand that particular word?
- Try to identify the word family. Part of the word may well be related to another English word which you already know. Look out for prefixes, suffixes and compounds, e.g. disapproval = approve + prefix & suffix.
- Many English words are of French or German origin, so think about whether there is a similar word in your own language or other foreign languages you know, e.g. discipline - Disziplin.
- Does the sound of the word help? Say the word out loud, e.g. "Ping!".

File 4 Using an English-English dictionary

A good English-English (monolingual) dictionary provides you with:

- the different forms of a word.
- the class of the word (noun, verb, adverb, etc.).
- the pronunciation (phonetics).
- the different meanings.
- grammatical information.
- examples showing how the word is used in context.
- the prepositions, etc. commonly used with the word.
- idioms, etc. in which the word is commonly used.

Tips for using an English-English dictionary

- Before looking up a word identify whether it is a verb, noun, adjective, adverb, etc.
- When you have found the entry for the word make sure you look at the meaning for the correct class of word.
- Read any additional information which the dictionary gives you about the word.
- Take your time to read the entry carefully. Make sure you find the entry which matches the use of the word in the given context.

File 5 Writing a composition

A composition is a clearly structured piece of writing in which important points on a topic are presented. Compositions are generally written in the present tense.
In an argumentative composition you express your opinion and give arguments for and/or against a given topic.

Stage 1: Preparation

- Define the topic clearly.
- Brainstorm ideas and write down what you know about the topic in a mind map, diagram or table.
- If necessary, collect facts, arguments, statistics and other information from other sources, e.g. your library or the internet.
- Formulate your opinion on the topic.

Stage 2: Arranging your material

- Decide on the most important facts, arguments or pieces of information that you want to include in your **composition** (maximum 5 items).
- For each main argument try to think of two supporting arguments.
- Put your facts or arguments in a logical order. For example, you could:
  - start with general ideas and then go on to talk about details.
  - put things into chronological order.
  - start by giving all the arguments in favour and then give all the counter-arguments.
  - follow each argument in favour with a counter-argument.

Stage 3: Writing your composition

The introductory paragraph

- The introductory paragraph should introduce your topic and catch the reader's interest. You could:
  - give a relevant quotation.
  - make a provocative statement.
  - describe a relevant (humorous) personal experience.
- In your introductory paragraph you should also:
  - say clearly what your **composition** is going to be about in one or two sentences.
  - ask a question or questions which you will answer in your **composition**.
  - give a brief outline of the problem/topic at present and in the past.

The main paragraphs

- Start a new paragraph for each new idea.
- Begin each paragraph with a main idea or argument and then follow it up with your supporting or counter- arguments (see also Stage 2).
- Write simple sentences.
- Arrange your sentences in a logical order.
- Join your sentences using suitable connectives → SF 7
- Use adverbs to emphasize and tone down points.
- Use examples where possible.

The conclusion

- Do not introduce new ideas in your conclusion!
- Make a brief concluding statement summarizing what you have said in your **composition**.
- Include a quotation or anecdote if you can.

Stage 4: Checking your composition

- Check that all your sentences are complete and that they make sense.
- 'Proofread' your **composition**: check for grammar, spelling and punctuation mistakes.
- Write or print out the final version of your composition.

File 6 Writing a summary

A summary is a short piece of writing in which you restate the main points of a text in your own words.

- A summary is almost always written in the present tense.
- Use your own words and do not copy any phrases or sentences from the text.
- Do not use statistics, numbers, figures, details or quotations from the text.
- Do not give your personal opinion.
- Your final version should be no longer than between one fifth and one third of the length of the original.

A summary of a non-fictional text

In a summary of a non-fictional text, for example, a newspaper article, a report, a factual text, etc. restate the main points or arguments presented in the text.

1. Read through the original text at least twice.
2. In your own words make a note of the main arguments or points in the text. You could:
   - Make a list of key words.
   - Write one sentence to go with each paragraph.
3. Rewrite some or all of your sentences to say what the writer's intention was. You can use the following phrases:
   The author/writer…
   - points out that …
   - emphasizes …
   - suggests that …
   - describes (how) …
   - concludes by saying that …
4. Link your sentences with connectives → SF 7

A summary of a fictional text

In a summary of a fictional text, for example, a short story or extract from a novel, describe briefly what happens in the story or extract and who the main characters are.

1. Briefly describe the characters and the setting.
2. Briefly describe what happens and how the events are connected. Use these phrases:
   The scene/story/extract …
   - … takes place in …
   - … is set in …
   - … takes place in the present/past/future.

   The main character is …
   There are two/three/etc. main characters.
3. Link your sentences with connectives → SF 7

File 7 Connectives

Connectives are words and phrases which you use to link sentences and paragraphs. Using connectives helps readers to follow your ideas more easily because they give your text a clear structure.
Make use of the following connectives in your texts. Keep your own list of personal favourites and add new words and phrases to it when you learn them.

To express a certain order:
to begin with • initially • first of all • first(ly)..., (second(ly)..., thirdly..., etc.) • next • then • later • finally

To add ideas:
also • as well • in addition • additionally • too • moreover • furthermore

To give examples:
for example • for instance • by way of illustration

To generalize:
as a rule • generally • in general • by and large • on the whole

to express certainty:
clearly • evidently • obviously • undoubtedly

To give a reason:
consequently • that is why • therefore • this is the reason why • as a result

To give an opinion:
in my opinion • to my mind • the way I see it • in my view

To compare or give an opposing opinion:
in the same way • similarly • likewise • equally • on the one hand ... on the other hand • in contrast • on the contrary

To conclude:
finally • in conclusion • to conclude • in summary • to sum up

File 8 Collecting words

Where to write new words

- On cards with the word in English on the front and the definition in English (and the German translation) on the back.
- In a vocabulary book: use a small ringbinder so that you can add pages and move pages around as necessary.
- On the computer.

Don't just write German translations to go with new English words. Always write an explanation or give a synonym or opposite in English!

Ways of collecting and organizing words

- Alphabetically (A – Z): This is good as a reference system and you have plenty of space for definitions. You can find words easily, but you have to know what the word is in the first place. It is a good method to use in addition to the other methods listed below.
- By writing synonyms and opposites: Group together words with similar meanings, e.g. pretty, nice, lovely, or with their opposites, e.g. pretty ≠ ugly.
- In word families: When you note a new word, include the other words which can be formed from it, e.g. the verb, adjective, adverb, etc. Also make a note of the word with its prefix or suffix (see opposite). This is a good way of quickly increasing your word power!
- In collocations: Make a note of words which are often written together (verb + preposition, verb + noun, etc.). Always try to learn words with their collocations, e.g. to make allowances.
- In semantic fields: Group words according to topics. Include nouns, adjectives, verbs, idioms, phrases, etc., e.g. School: to attend, to be absent from, to play hockey; classroom, hall, staff room.
- In mind maps by topic: Start with the topic heading and add all the words which come to mind in a logical way.

Be flexible: choose the method for collecting and organizing words which best suits your needs at a particular time!

Prefixes

A prefix is added to the beginning of a word to change its meaning.

To give a word a **negative** or **opposite** (≠) meaning add:

Prefix	Example
in-	active ≠ **in**active
im-	possible ≠ **im**possible
il-	literate ≠ **il**literate
ir-	relevant ≠ **ir**elevant
un-	happy ≠ **un**happy
dis-	to like ≠ to **dis**like
de-	to stabilise ≠ to **de**stabilise

To show an **increase** (+) or **decrease** (–):

Prefix	Example
over- (+)	to heat – **over**heat
out- (+)	to grow – to **out**grow
super- (++)	natural – **super**natural
under- (–)	to pay – to **under**pay
sub- (–)	standard – **sub**standard

Others:

Prefix	Meaning	Example
inter-	between/among	**inter**national
pre-	before	**pre**-school
post-	after	**post**-war
trans-	across	**trans**atlantic
counter-	against	**counter**revolution
pro-	for	**pro**-abortion
anti-	against	**anti**-smoking
re-	again/back	**re**read
mis-	wrong/bad	to **mis**understand

Suffixes

A suffix is added to the end of a word to give it a different meaning, usually to make a verb or adjective into a noun:

Suffix	Example
-ence	to differ – differ**ence**
-ment	to improve – improve**ment**
-ion/-ation/-ision	to nationalize – nationaliz**ation**
-er/-or	to teach – teach**er** / to act – act**or**
-ing	to begin – begin**ning**
-ness-	happy – happi**ness**
-(i)ty	difficult – difficul**ty**
-th	true – tru**th**

File 9 Mind maps

Mind maps are useful for collecting and organizing a large amount of information about a certain topic. You can make a mind map for yourself or as a group. The basic structure of a mind map is always the same.

How to make a mind map

Use a pencil so that you can easily delete things or change their position, if necessary.

1. Write your main topic in the middle of a page and draw a circle around the word or phrase. (Fig. 1)
2. Think of important things directly related to the main topic. Write each idea (subtopic) in a circle around the outside of your first circle. Draw lines connecting the subtopics to the main topic. (Fig. 2)
3. Now add more circles with related ideas around the subtopics. (Fig. 3)

Instead of drawing circles you can use lines with 'branches' for subtopics. (Fig. 4)

Ways of using mind maps

- To brainstorm ideas on a given topic:
  Write the topic heading in the centre circle and add all new and related aspects.
- To organize information from a text (or other source):
  Start with the most important point in the text, add the main subtopics, then add relevant detailed information or quotations from the text.
- To develop word fields:
  Start with the generic term and add all related words.
- To organize arguments:
  Start with the topic of discussion. Add a branch or circle: 'for' and another one 'against'. Continue by adding further branches/circles with arguments 'for' and 'against.

Fig. 1

Fig. 2

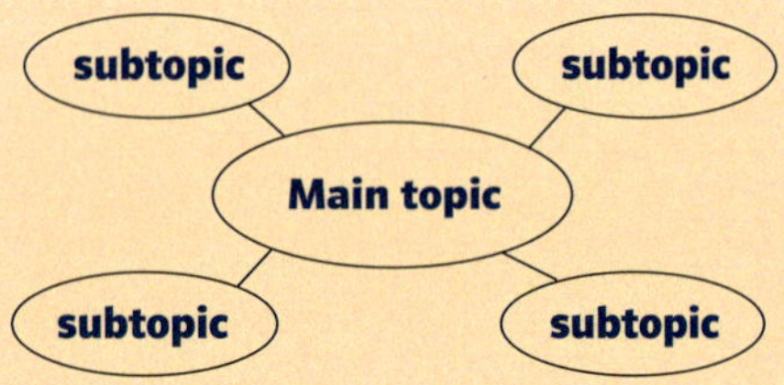

Fig. 3

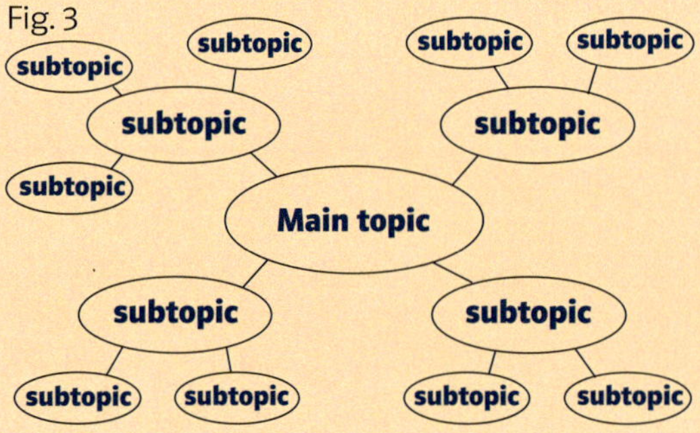

Fig. 4

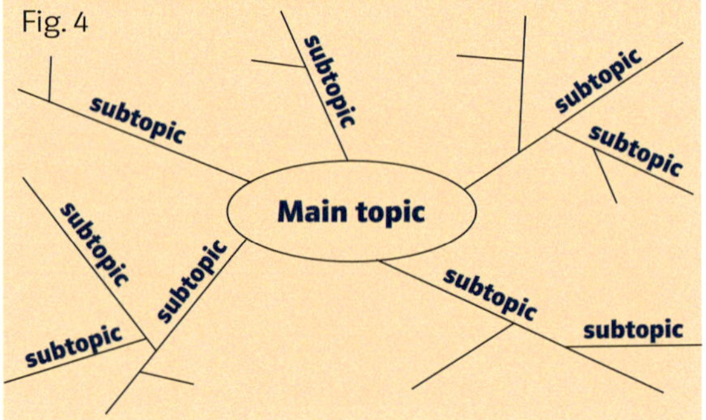

File 10 Translating/Mediation

Guidelines

1. Before you start translating
   - Always read the whole text before you start translating.
   - Make a note of the main points in your own words in German before you start your detailed translation.
2. First version
   - Write your first version in pencil so that you can change and correct things easily.
   - Don't translate word for word, but do keep as close to the original as possible.
   - Don't leave any gaps! In your first version write words or phrases which you can't translate in English.
3. Second version
   - Read through your first version and try to translate any remaining English words from the context. If you can't, look them up in a dictionary.
   - Check your translation word for word against the English text and make sure you haven't left anything out. Make corrections as necessary.
4. Final version
   - Reread your text and decide if it sounds German. If it doesn't, you haven't translated properly. Think about what you are trying to say and decide how you would say it in German. Make corrections.
   - Reread your translation and repeat these steps until you are certain your text is in good German. If you have time, leave the text for a while (e.g. overnight) before checking it for the last time.

Mediation

1. Follow the six steps that are mentioned in Skills File 1.
2. Summarize the main ideas e.g. of a paragraph or of a complex sentence in German.
3 Note: A mediation is **not** a translation! Do not translate sentences or words literally. Your task is to outline the meaning of the specific section.
4. Check if your expressions, your style and your word order are appropriate in German.

Tips and tricks

1. Don't improve texts when you translate. For example, if one verb is used repeatedly in the text use one verb in your translation, even if you think another would sound better.
2. When translating long sentences, define the main clause and translate this first, then translate the other clauses in their order of importance.
3. Try to keep to the same number of sentences in the text and in your translation. (Don't split sentences up or join them in your translation, unless it is really necessary.)
4. Be prepared for your German text to be longer than the English original: English sentences are often shorter than German ones. One of the reasons for this is that, for example, there is often a relative clause in German where in English a participle construction (with -ing form or past participle) is used.

Common 'false friends'

actual	eigentlich (≠aktuell)
after all	doch, zum guten Schluss (≠nach allem)
brave	mutig (≠brav)
briefcase	Aktentasche (≠Brieftasche)
critic	Kritiker (≠Kritik)
gift	Geschenk (≠Gift)
gymnasium	Turnhalle (≠Gymnasium)
however	jedoch (≠wie auch immer)
to overhear	zufällig mitbekommen (≠überhören)
physician	Arzt (≠Physiker)
rent	Miete (≠Rente)
rubber hose	Gartenschlauch (≠Gummihose)
sensible	vernünftig (≠sensibel)

Translating/Mediation (contd.)

Common problems

Problem/Solution	English	German
Present perfect sentences with ‚since/for' are usually present tense in German.	I have had my car for 2 years now.	Ich habe mein Auto jetzt seit zwei Jahren.
Constructions using past participles and -ing forms can often be translated with a relative clause.	I thanked them for the lovely weekend spent with them. A group of American students visiting our town were at college today.	Ich dankte ihnen für das schöne Wochenende, das ich mit ihnen verbracht hatte. Eine Gruppe amerikanischer Studentinnen, die unsere Stadt besuchen, waren heute in der Schule.
The translation of 'it' is not always 'es'. It depends what 'it' refers to.	I couldn't get a seat on the train: it was totally crowded.	Ich konnte im Zug keinen Platz kriegen: er war total überfüllt.
'you' and 'they' can often be translated with the German 'man'.	In Britain you often see double decker buses.	In Großbritannien sieht man oft Doppeldeckerbusse.
'do' or 'did' in positive sentences are used to emphasize the verb. In German you usually need an adverb.	I did do my homework!	Ich habe sehr wohl meine Hausaufgaben gemacht!
'there is/are' and 'there was/were' are usually translated with 'es gibt' or 'es gab'.	There's so much to do and see in Atlanta! There was nothing to eat.	Es gibt so viel zu tun und zu sehen in Atlanta! Es gab nichts zu essen.
An adjective + noun combination in English can often be translated by one noun in German.	congressional elections (adjective + noun)	Kongresswahlen (compound noun)
'with + -ing form' is best translated with a separate clause beginning with a conjunction (weil, da, wo, wenn, wobei, etc.).	Roads are becoming busier and busier with more and more people driving cars. With winter approaching, they decided to leave.	Die Straßen werden immer voller, weil immer mehr Leute Auto fahren. Jetzt, wo der Winter sich näherte, beschlossen sie zu gehen.
Passive sentences can often be translated in the active form using 'man'.	The stolen money was found in a phone box.	Man hat das gestohlene Geld in einer Telefonzelle gefunden.
Translate proper names etc. only when there is a real German equivalent.	Henry VIII Cologne Prince Charles	Heinrich der VIII. Köln Prince Charles
Don't convert miles to kilometres, pounds to kilogrammes, etc.	She runs 5 miles every day.	Sie läuft jeden Tag 5 Meilen.

File 11 Letter writing: Formal letters & commercial correspondence

A formal letter can be a letter of application, a reservation, a request for information, etc.
Commercial correspondence refers to formal letters written in a business context, i.e. from one company to another. There are different ways of layouting these types of letters. The model letter below is in 'block form', which is probably the best to use because it is accepted everywhere.
The parts in blue are only necessary in commercial correspondence.

Punctuation: In 'block style' you don't use commas in the address, salutation or complimentary close.

Writer's address. In commercial correspondence you should include the company name, telephone, e-mail, Internet address and fax number.

Weltweit Services
Vinckestr. 1
70183 Stuttgart
Germany

Tel: 0711 768232
Fax: 0711 768233
e/mail>
www. ...

Addressee, including name and position, if known.

Personnel Manager
Import Export
11270 White Blvd.
Los Angeles CA 91430
USA

In commercial correspondence include references. 'Our ref:' is usually the initials of the person who wrote the letter. (TW = Thomas Wagner)

Your ref:
Our ref: TW

Date: The British write day, month, year (11.1.00). Americans write month, day, year (1.11.00). So it is best to write the date in full to avoid confusion.

11 January 2000

Salutation:

Dear Mr ... / Ms ...

In commercial correspondence include a subject line beginning 'Re:'

Re: Your advertisement in the Sun Times of April 28

Main part of letter, in which you state your reason for writing in formal, polite language. Group your points in paragraphs.

With reference to your advertisement in the Sun Times of April 28, I would like to apply for the position of bilingual buying assistant.

I am particularly interested in this job as I would very much like to gain work experience abroad. As well as speaking fluent English and Spanish, I have a working knowledge of French. My mother tongue is of course German.

I can supply references from former employers if you wish.

I enclose my curriculum vitae. Please do not hesitate to contact me if you should require further information.

Complimentary close:

Yours sincerely

Signature

Thomas Wagner

Name in full

Thomas Wagner

In commercial correspondence: Writer's position in company.

Trainee buyer

Enc.

Enc. = Enclosures: indicates that additional material is being sent with the letter.

File 12 Letter writing: Personal letters

A personal letter is a letter you write to someone you know well on a personal basis, e.g. a friend or relative. There are no special rules which you have to follow when you write to people you know well. The model letter below shows you what an informal letter generally looks like in English. You can use informal language in a personal letter, as if you were speaking to the person.

Your address goes in the top right hand corner.

Vinkelstr. 1
70183 Stuttgart
Germany

Write the date under the address.

May 3, 200...

Start the letter: Dear + name,

Dear Frank,

Use a capital letter for the first word.

Just a few lines to let you know I'm still alive! Sorry for not having written for so long, but I've been very busy at college with exams.

Group your news, etc. in paragraphs.

How are things with you? What are you doing for the summer? Unfortunately my applications for jobs in the States have all been turned down so far. Oh well, I'll keep trying...

Please drop me a line when you have a minute. We really must keep in touch. Give my regards to your family.

Finish the letter with one of these phrases: All the best, Best wishes, Bye for now, Love, Lots of love,

All the best,

Thomas

P.S. If I do get to the States, I'll let you know my address. Maybe we'll be able to meet up!

Salutations & complimentary closes

The salutation you choose in a formal letter depends on whether you know a person's name, sex, etc.
The complimentary close usually depends on the salutation you have used, but variations are possible. Follow the guidelines below.
Americans usually close formal letters with: Sincerely yours or Yours truly

Salutation	When used	Complimentary close
Dear Sir or Madam	to a company and when you don't know person's name or sex	Yours faithfully
Dear Madam	to a woman, when you don't know her name	Yours faithfully
Dear Sir	to a man, when you don't know his name	Yours faithfully
Dear Mr Brown	to a man, whose name you know	Yours sincerely
Dear Ms Brown	to a woman, whose name you know	Yours sincerely
Dear Jenny	to a person you know well	Best wishes/Yours

Salutation: the way a person is addressed at the beginning of a letter

Complimentary close: the phrase used before signing off

File 13 Giving a talk or presentation

Preparation

1. Gather material on your topic.
   - Brainstorm to find ideas on the topic and make a mind map.
   - Find information from relevant sources (library, school books, the internet, newspapers, etc.)
2. Choose which main points you want to talk about.
3. Make a set of prompt cards in English, which you can look at during your talk to remind you of what you want to say. (See below.)
   - Write down the key words, phrases, and vocabulary which you will need.
   - Note down any examples or quotations.
   - Note down any visual aids (OHPs, photographs, posters, flip charts, etc.) which you want to use.
4. Arrange your prompt cards in a logical order. You could present points:
   - in terms of importance, from the most to the least important (or vice versa).
   - chronologically.
   - in favour and then all those against.
5. Make a prompt card for your introduction (see below). To catch the listeners' attention you could start with:
   - an interesting question or fact
   - a quote or joke
   - a picture or cartoon on an OHP transparency.
6. Make a prompt card for your conclusion (see below).
   - Note the point(s) you want to emphasize.
   - Write down key words and phrases.
   - Make a note of a final quotation or example to use.

Giving a talk or presentation

Practise your talk or presentation out loud preferably in front of a mirror or ask a friend/relative to listen to you!

1. Speak slowly and clearly!
2. Use short, clear sentences.
3. Vary your voice so that you don't sound boring. For example, emphasize particularly important points by speaking a little more loudly.
4. Look at your audience as often as possible. Don't look at your prompt cards all the time.
5. Try not to make distracting movements such as waving your hands, etc.
6. Smile! (But don't laugh nervously!)
7. Start by introducing yourself (if necessary) and your topic. Finish by thanking your audience for listening and asking if anyone has any questions.

Useful phrases

Good morning/afternoon, my name is …
Today I'm going to talk about …
The subject of my talk today is …
Can everyone hear me alright?
Can everyone see the board/screen/etc.?
I'd like to begin by (+ verb in -ing form)
Let me start by saying a few words about …
Let me give you an example.
As you can see on the chart/graph/cartoon/etc. …
I'd like to finish/conclude by (+ verb in -ing form)
Let me finish by (+ verb in -ing form)
Finally …
Thank you very much for your attention.
I hope you've all enjoyed my talk.
I'd be happy to answer any questions.

Prompt cards

Introduction
– note on main points of talk
– key words for example/quote/joke
– vocab & useful phrases

Point 1
– key words
– example/visual material
– vocab & useful phrases

Conclusion
– note on main points
– key words for quote/joke
– vocab & useful phrases

File 14 Having a discussion

- Use the phrases below.
- Always be polite!
- Look at the people you are talking to but don't stare at them.
- Don't interrupt unless you feel it is really necessary.
- Listen carefully to what other people are saying.
- Keep to the point.
- Don't talk too much.

Leading a discussion

A discussion leader:

- introduces the discussion and the participants
  - (As you know,) our discussion today is about …
  - On my right/left I have … (name of person or group), who is/are in favour of/against …
- makes sure the discussion is held in an orderly way and that participants are always polite to each other.
- chooses whose turn it is to speak.
  - Who would like to open the discussion?
  - … , would you like to open the discussion?
  - The next person to speak is … , followed by … and then … .
- makes sure everyone has an equal chance to speak, including people who are shy.
  Is there anything you would like to say, … ?
- makes sure no one person monopolizes the discussion.
  - I'm sorry to interrupt you, but I'm afraid you'll have to stop there.
  - Sorry, but I must give the others a chance to put forward their arguments now.
- makes sure that only one person is speaking at a time.
  - Please wait your turn, you'll have a chance to speak in a minute!
  - If you'd just let … finish, please.
- notes down the most important points and summarizes the discussion at the end.
  - Time is running out, so I'd like to sum up what we've been discussing.
  - We heard from … that …
- always remains neutral (i.e. doesn't give his/her own opinion).
- thanks everybody for taking part at the end.
  - Thank you for taking part in the discussion!

Useful phrases for discussions

To make a suggestion:

Let's consider …
I suggest/propose that you/we …
I suggest/propose + (verb in -ing form)
What do you think about … ?
If I were you, I would …

To respond to a suggestion:

That's a good idea!/Fine!
OK/Yes, why not?
OK, if you want, I don't really mind.
I think it would be a better idea to talk about/consider ….
Couldn't we look at this/do this another way?
What about … ?
I'm not sure (if that's such a good idea). How about … ?

To ask for an opinion:

What do you think/feel about … ?
What is your view/opinion on/about … ?
What would you say?

To give an opinion:

As far as I'm concerned …
I think/feel/believe/am of the opinion that …
To my mind …
In my opinion …/In my view …

To agree:
You are (quite/absolutely) right.
That's a very good point!
Absolutely!/Exactly!
I agree (with you/Susi/etc.).
You have a point.
You might/could be right.
I think you're right to a certain extent/up to a point.

To interrupt someone:
Sorry to interrupt, but (could you please explain/ repeat the point you just made).
Excuse me for interrupting, but (I don't understand what you just said).
Sorry for interrupting but (can I please say something/ make a point?)

To change the subject:
There is something else that I'd like to say …
Have you ever thought of/considered …
On the other hand …
Try looking at things another way/from a different angle.
I'd like to bring up another point/point out something else.

To ask for further explanation:
Did I get that right?
Do you really mean that … ?
Did I understand correctly that you think … ?
Could you explain what you just said/the point you just made/etc. again, please?
Could you give me/us a concrete example, please?

To disagree:
I disagree (entirely/with some of the things you said/etc.)
I'm sorry, but I completely disagree.
Stop exaggerating!
I think you're oversimplifying things.
That's an interesting point, but …
That's not the point.
You've missed the point (entirely).

To deal with being interrupted:
Can I just finish what I was saying, please?
Just a second. I haven't quite finished.
Would you let me/allow me to make my point, please.
You could at least let me finish my sentence!

To return to the original subject:
Lets get back to what we were saying/discussing.
Let's get back to the point.
As I was saying before, …
Getting back to what we/you/… said earlier …

To say that a point has been misunderstood:
I'm sorry, but …
… that's not what I meant to say (at all)!
… I think you've got the wrong end of the stick!
… you've got me (completely) wrong.
… you've (completely) misunderstood me.
… that's not what I was trying to say!
… I didn't make myself clear. I'll try and explain again.

File 15 Small talk

Small talk is light (casual) conversation. It is a way of showing interest in others, communicating in a friendly way and avoiding conflict at all times! At a social event, like a party, you might spend the whole time in casual conversation. At a business meeting, on the phone, at an interview, etc. you can start (and end) with a few minutes of small talk.

(especially popular in Britain!)

- A journey (to or from the place where you are)
- Where you live./Where your conversation partner lives
- Your family and /or your conversation partner's family (if you know them)
- A recent/well-known TV programme or film
- Tips about what to see and do in a certain town or city (i.e. cultural events, attractions, restaurants, local specialities, etc.)
- Sport (e.g. a current international sports event)
- A vacation
- Where you/your conversation partner went to school or university
- What you/your conversation partner do for a living
- Hobbies and interests
- An experience/event that you and your conversation partner are both currently involved in or have been involved in
- A non-controversial item in the news

Useful phrases

Starting a conversation
Hi!/Hello!
What are you up to?
Do you mind if I join you?
Nice to see you (again)!
How are you?
How are you doing?
How are things?

Filling time
Well, …
Actually, …
Let's see, …
Let me think, …

Finding a topic
Did you see … on TV last night/at the weekend/etc.?
Have you seen the new [James Bond] film yet?
Did you read about [that terrible train crash] in the newspaper this morning?
Terrible/Beautiful weather, isn't it?
How was your journey?
How did you get here?
What made you decide to come here?
Have you ever been here before?
Who do you think is going to win [the football match] this evening?

Responding & showing surprise
Really?
I don't believe it!
That's incredible!
Well, I've never heard of anything like that before!
That's the funniest/most interesting/strangest thing I've heard in ages!
I've never thought of it like that before.
I couldn't agree more!
That's funny! Exactly the same thing happened to me the other day/two weeks ago/etc.

Using question tags
You know …, don't you?
He's … , isn't he?
It hasn't …, has it?
(See grammar file on p. 182.)

Finishing a conversation
It was nice talking to you!
Goodness/Oh no, is that the time?
Excuse me,/What a shame, I've got to go now.
Hope to see you again soon.
Hope you get home safely.
Take care.
Give my regards to … .
Don't forget to say hello to … for me!

File 16 Making phone calls

Basic guidelines

Book room for
3 nights from
27th July.
How much?
Evening meals?

- Prepare yourself thoroughly.
  - Make notes in English or write out a rough draft of the main things you want to say or find out to give you more confidence and fluency. This is especially important when making formal or business calls.
  - If possible find out the name and number of a direct contact person in advance.
  - Think of things the contact person is likely to ask or say and work out responses.
- Speak clearly and slowly. State your requests clearly and to the point.
- Don't expect to understand every single word that is said.
- Don't worry about making mistakes.
- Be polite! Remember to say 'please' whenever you ask for information and 'thank you' whenever you have been given help. Remember:
  please = bitte • you're welcome/not at all = bitte schön; gern geschehen
- State telephone numbers as single digits. When the same number occurs twice consecutively you can say 'double' (e.g. 33: double 3). 0 = Oh (BE & AE) or zero (AE)
  Example: 07145 14406: Oh/zero - seven - one - four - five - one - double four - oh/zero - six
- When making a social call, give your first name only.
  When making a formal/business call give your first name and surname.
  Address people you know well (friends, people you know socially, people of your own age) by their first names.
  In a formal/business situation address people using 'Mr', 'Mrs' or 'Ms' [mz] + surname.
- Be prepared to spell words, especially people's names and names of places. Useful phrases:
  Would you like me to spell .../that for you? It's spelt ... • I'll just spell that for you. It's ...
- Ask the caller to spell names, etc. for you to avoid confusion. Useful phrases:
  Could you spell .../that for me, please? • Would you mind spelling .../that for me, please?
- Repeat information (times, names, addresses, prices, etc.) to be sure that you have understood everything correctly. Useful phrases:
  Let me just check that I've understood all the information you've given me/asked for correctly. ...
  I'll just repeat that back to you to make sure I've got everything down correctly. ...
  Let me just check that I've spelt your name correctly. ...
  Let me just check that I've written the dates/times/etc. down correctly. ...
- If you don't understand something:
  - Ask the person to repeat what they have just said: Could you repeat that, please?
  - Ask the person to explain something in a different way:
    I'm sorry, I'm afraid I don't understand what you mean. Could you explain that again, please?
  - Ask the person to speak more slowly: Could you speak a little more slowly, please?
- If you can't hear the caller properly, ask the caller to speak more loudly:
  I'm sorry, (it's a very bad line,) I didn't quite catch that! Could you speak up a bit, please?
- If you think you've got the wrong number, check and apologize. Don't just hang up. Useful phrases:
  Is that (+ number you want) ? • I'm sorry, I think I've dialled the wrong number, I'm trying to call (+ number you want). • I'm sorry to have bothered you.
- If the caller has got the wrong number, explain politely. Don't just hang up.
  I'm sorry, but I think you've got the wrong number. This is (+ your number/name/company name). • I'm sorry, there's no-one here with that name. This is (+ your number/name/company name).
- Don't interrupt! If you do interrupt, be as polite as possible. There are examples of phrases you can use to interrupt someone or deal with being interrupted in Skills File 14 .

File 17 Social calls

■ Remember to follow the basic guidelines in Skills Files 16. (See also ideas and useful phrases in Skills File 15)

Greeting
Hello! • Hello, this is…[1]. • Hello …[2].! It's me, …[1].

Starting a conversation
How are you? • How are you doing? • It's great to hear from you! • I just thought I'd call and find out how you are. • I just wanted to ring and let you know that I'll be in the States in July/etc.

Asking to speak to someone else & passing a call on
Could I speak to …[2], please? • Is …[2] in, please? • Do you know when …[2] will be back, please?
Just a second, I'll go and find him/her. • Hold on a minute, I'll see if he/she's in.

Giving and taking messages
Could you give …[2] a message, please?
I'm afraid he/she can't come to the phone right now. Can I take a message? • I'm sorry, but he/she's not in at the moment. Would you like me to take a message?

Ringing back
Would it be OK to ring back later? • What time can I ring back till? • I'll ring back in about half an hour, if that's OK. • Could you give …[2] my number, please, and ask him/her to ring me back? • Would you ask …[2] to ring me back, please?
You can try ringing again in about 20 minutes if you want to. • He/she should be back at about 10 pm. • Shall I ask him/her to call you back?

Ending a call & saying goodbye
Thanks for ringing! • It was great to talk to you/to hear from you! • Speak to you again soon, I hope. • Bye! • Take care. • All the best.

1 Your name.
2 Name of person you want to speak to.
3 State telephone/extension number.

File 18 Formal phone calls

■ Remember to follow the basic guidelines in Skills Files 16.

Greeting
Hello/Good morning./Good afternoon. …*.
(This is) …[1] speaking. • Can I help you? • What can I do for you?

Asking to speak to someone
Can I speak to …[2], please? • Can you put me through to …[2], please? • I'd like to speak to someone in your … department.

Holding the line & putting calls through
Please hold (the line). • One moment, please. •
I'll connect you with …[2]./with the … department. •
I'll put you through (to …[2]/to the … department).

When someone isn't available …
I'm sorry, …[2] isn't here/available at the moment/ today/etc.

Getting the right extension, etc.
Can you give me …[2] 's extension number, please? •
I'd like to receive more information about … . •
I'm calling about … . • I can give you his/her extension number/voice mail number/fax number/ e-mail address. It's …[3].

Giving and taking messages
Would you like to leave a message? • Can I take a message? • Would you like me to pass on a message? • Could you pass on a message, please?

Ringing back
Shall I ask him/her to call you back? • Could you call back later/in half an hour/etc., please? • Could you ask him/her to call me back, please? My number is …[3] • Can I call back later? • I'll call back later/in half an hour/etc. , if that's OK. • When would be the best time to call back?

Ending a call
Thanks for your call!/Thanks for calling! • Thank you very much for your help! • Sorry I can't be of any more help. • Have a nice day! • Goodbye!

1 Your name.
2 Name of person you want to speak to.
3 State telephone/extension number.
* Name of company.

File 19 Describing illustrations & photographs

Types of pictures

photo(graph) • print • transparency • slide • painting • drawing • sketch • picture • illustration • image • collage • electronic image • snapshot • skyline • panorama • close-up • action photo • still (from a film)

Saying what a picture is of

This is a photo/an illustration of … .
This photo was taken in … .
The illustration shows …
In this illustration/photo you can see (that) …

Saying where things are in a picture

on the right-hand/left-hand side of the picture • in the centre • in the foreground · in the background • at the top • at the bottom • in the upper right-hand corner • in the lower left-hand corner

Adjectives of style & colour

colour • black & white • bright/dull • hand-painted/hand-drawn/drawn by hand • clear/in focus • blurred/not in focus/out of focus • lifelike • to scale/not to scale • posed/staged

Describing the impression a picture makes

- The person in the picture looks … (+ adjective, e.g happy)
- The photo/illustration supports the point made in the text that …
- The photo/illustration gives you the impression that …
- The impression I get from this photo is that …
- The illustration makes me feel … (+ adjective: e.g. worried).
- The colours the artist has chosen make me feel … (+ adjective: happy/sad/worried/etc.)
- In my opinion the illustration would be better, if … (+ simple past: e.g.if it was more life-like)

File 20 Describing cartoons

Cartoons are funny but critical. When looking at a cartoon the most important thing is to try to work out what point the cartoonist is making. Generally the point is made by a combination of the text and the illustration, so it is important to analyze and understand both.

Illustrations generally show caricatures of well-known people or particular stereotypes. To understand a cartoon, you need to recognize and define the person or group of people illustrated.

Text can be in one or all of the forms below. All text is relevant!

- A caption (a statement in quotation marks under the illustration). You need to work out who is speaking.
- A speech bubble, or speech bubbles.
- Other text, for example, on a sign or poster illustrated in the cartoon.

Phrases for talking about cartoons

Describing the cartoon

The cartoon consists of an illustration of … .
The illustration shows …
There is a caption under the cartoon, which says "…".
In the (first) speech bubble it says "…".
The text in the speech bubble is spoken by …
The figure is a stereotype/ caricature of …
The drawing is detailed/clear/sketchy/abstract.

Talking about the point

The cartoonist is making fun of …
The cartoonist is satirizing …
The cartoon is funny because of the misunderstanding between … .

Giving your opinion

I think/I don't think the cartoon is (very) funny because …
I get the joke/point. /I don't get the joke/point.
I think/I don't think the cartoon is easy to understand because …
I agree/I don't agree with the point the cartoonist is making because …

File 21 Describing statistics

Useful words and phrases

To describe an increase

to rise • to increase • to have increased by • growth • an increase (in the number of …)

To describe a decrease

to decrease • to drop • to decline • a decrease • a decline • a drop (in the number of …)

Useful adjectives and adverbs

sharp(ly) • slow(ly) • steady/steadily • continuing/ continual(ly) • gradual(ly) • significant(ly) • insignificant(ly) • negligible/negligibly • large • high • small

To express numbers & figures

the figures for the last year/the last month/etc. • the latest figures • the total number of … • a (significant/insignificant/etc.) number of … • a (high/etc.) percentage of … • the statistics • a majority/minority of …

Useful verbs

to show that … • to indicate that … • to suggest that … • to give information about … • to make it clear that

Types of graphs

graph • chart • pie chart • bar chart • diagram

Types of surveys

survey • opinion poll • questionnaire

Qualifiers

a total of • over • under • nearly • more/less than • between • the same number as • the same amount of • from over/under … to just over/under • half • a third • two thirds • a quarter • (25) per cent • twice as many • three times as many • on average • averagely • roughly • exactly • precisely

File 22 Projects & group work

Finding information

Set up an 'information centre' in your classroom (or of your own) in which you keep a record of where to find information. Find and keep a record of:

- the name, address, e-mail address, telephone and fax number of:
  - your local library.
  - the nearest university or college library.
  - your twin-town association (if your area has one).
  - the nearest newsagent or bookshop which stocks English newspapers and magazines. (Tip: Try your nearest main station.)
- the website addresses of:
  - search engines in America and the UK.
  - schools and colleges in America and the UK.
  - youth organizations in America and the UK.
  - political parties in America and the UK.
  - newspapers, magazines in America and the UK.
  - government agencies in America and the UK.
- the names of any members of your class who have contacts (penfriends, relatives, friends, etc.) in the English speaking world, and who might be willing to help you.
- the names of any members of your class or in your neighbourhood who have an English-speaking parent/parents and who might be willing to help you.

Collect brochures from travel agencies and tourist information bureaus.

Collect material while you are on holiday (leaflets, magazines, postcards, advertisements, menus, timetables, etc.).

Add information and addresses to your files whenever you find new material.

Doing a project

How you go about doing a project will depend on the assignment. These tips are very general.

1. Read the information you are given about the assignment carefully.
2. In your groups, brainstorm to find ideas about:
   - what you all already know about the topic.
   - which aspects you want to concentrate on.
   - where you can find information.
3. If necessary, decide which points are more important (concentrate on these) and which are less important (and can possibly be left out).
4. Decide which tasks each member of the group are going to do. Make sure the workload is divided equally amongst you.
5. If not specified in the assignment, decide how you want to present your material. You can produce:
   - a talk using visual aids.
   - a video.
   - a cassette or CD.
   - a leaflet.
   - a brochure.
   - a file of written information.
   - a poster or wall chart.
   - a written report.
6. Remember to keep a record of your sources, i.e. where you found material, where quotes or extracts from original material are taken from, etc.

Useful websites (general information)

Search engines

www.altavista.digital.com
www.excite.com
www.lycos.com
www.yahoo.com
www.google.com

GB

www.searchuk.com (Search engine for GB only)
www.bta.org.uk (British Tourist Association)
www.britcoun.org (British Council)

USA

www.usa.gov (US Information Agency)
www.whitehouse.gov (US Government)
www.nytimes.com (New York Times)

Grammar files

Indefinite article (Unbestimmter Artikel)

Form

a	college student university	vor **gesprochenen** Konsonanten
an	interview address hour	vor **gesprochenen** Vokalen

Verwendung

I have bought **a** book for you.	für zählbare Begriffe in der Einzahl
He is **a** teacher.	vor Berufsbezeichnungen

Definite article (Bestimmter Artikel)

Form

the	course language European	vor **gesprochenen** Konsonanten Aussprache [ðɛ]
the	engineer office hour	vor **gesprochenen** Vokalen Aussprache [ðiː]

Verwendung

the people of London **the** life of Lucy Jordan **the** University of Manchester	bei Substantiven (Personen, Dingen, Abstrakta), wenn sie näher bestimmt sind oder werden
the Americans	für die Gesamtheit von Personen
the United States	für Ländernamen im Plural
the Thames, **the** North Sea	für Flüsse und Meere
the Browns	für Eigennamen im Plural
On their tour they visited **the** church.	für die konkrete Bedeutung als Gebäude

Verwendung ohne Artikel

Der Artikel wird abweichend vom Deutschen **nicht** verwendet:	
nature, life, time, people	für abstrakte Begriffe und Sammelbezeichnungen, wenn sie nicht näher bestimmt sind oder werden
Trafalgar Square, Buckingham Palace	für Ortsbezeichnungen
Switzerland, Normandy, Lake Windermere	für Länder, Berge und Seen im Singular
church, school, college When does church begin?	für Gebäudebezeichnungen, wenn die normale Nutzung gemeint ist
Americans	für einzelne Angehörige einer Personengruppe

Plural (Mehrzahl)

Form

student college	student**s** college**s**	Die meisten Nomen bilden den Plural durch Anhängen von -s.
universit**y** cit**y**	universit**ies** cit**ies**	Konsonant + y wird zu -ies.
addre**ss** mat**ch**	address**es** match**es**	Nach Zischlauten (z. B. s, ss, ch, sh) wird -es angehängt.
man woman child life	**men** **women** **children** **lives**	Einige Nomen haben unregelmäßige Pluralformen.

Verwendung

information, weather, nature, knowledge, work, furniture	Einige Nomen treten nur im Singular auf.
trousers, jeans, glasses, clothes, thanks	Einige Nomen treten nur im Plural auf.
news, the United States, the United Nations, economics, mathematics, electronics ! No news **is** good news.	Einige Nomen treten nur in der Pluralform auf, das Verb steht jedoch im Singular.

Some important prepositions

Zeitliche Präpositionen

on 20th September **on** Saturday **on** weekdays	**at** six o'clock **at** the age of 16 **at** night **at** noon **at** midnight	**by** day
in 1992 **in** the morning **in** the evening		twelve minutes **to** five
		a quarter **past** six

Räumliche Präpositionen

go **to** the cinema go **to** college I've been **to** Scotland
at home **at** school **at** the supermarket

Präpositionen der Art und Weise

go **by** bus, car, bike
go **on** foot

Possessive case (Genitiv)

'*s*-Genitiv

Verwendung bei Personen, Tieren, Zeiten

Henry**'s** exam the baker**'s** (shop)	Nomen im Singular + **'s**
his parents**'** computer shop the Greens**'** house	Nomen im Plural + '
people**'s** faces children**'s** books	bei Nomen im Plural **ohne** s Anhängung von **'s**
today**'s** newspaper two weeks**'** pay	Bei Zeiten gelten die gleichen Regeln.

of-Genitiv

Verwendung bei Sachen

the title **of** the book the price **of** the computers	Der *of*-Genitiv wird im Singular und im Plural verwendet.

Comparison of adjectives (Steigerung von Adjektiven)

Form

cheap small	cheap**er** small**er**	cheap**est** small**est**	Steigerung der kurzen Adjektive mit -er und -est
nice	nic**er**	nic**est**	stummes -e entfällt
heavy	heav**ier**	heav**iest**	-y wird zu -ier und -iest
big	bi**gg**er	bi**gg**est	Nach kurzen Vokalen wird der Endkonsonant verdoppelt.
good bad little	**better** **worse** **less**	**best** **worst** **least**	Einige Adjektive haben unregelmäßige Steigerungsformen.
careful difficult	**more** careful **more** difficult	**most** careful **most** difficult	Alle dreisilbigen und mehrsilbigen Adjektive und alle zweisilbigen Adjektive, die nicht auf -er, -le, -y, -ow enden, werden mit *more* und *most* gesteigert.
careful difficult	**less** careful **less** difficult	**least** careful **least** difficult	Eine Verminderung der Adjektive ist mit *less* und *least* möglich.

Verwendung

In town centres walking is often **as fast as** going by car.	Gleichheiten werden mit *as … as* ausgedrückt.
Travelling by plane is **faster than** travelling by train. Travelling by car is **more expensive than** walking.	Komparative werden mit … *than* gebildet.
A bicycle is **not so expensive as** (isn't as expensive as) a car. A bicycle is **less expensive than** a car.	Eine andere Form Ungleichheit auszudrücken, ist *not so … as* oder umgangssprachlich *not as … as*. Wie im Deutschen kann man auch „weniger als" verwenden.
The Concorde was **the fastest** airliner from Paris to New York.	Der Superlativ wird nur benutzt, wenn mehr als zwei Dinge miteinander verglichen werden.

Adjective and adverb (Adjektiv und Adverb)

Form

Adjective	Adverb	
cheap lazy horrible economic fast good	cheap**ly** laz**ily** horrib**ly** econom**ically** **fast** **well**	Das Adverb wird gebildet aus Adjektiv + -ly. -y wird zu -ily -le wird zu -ly -ic wird zu -ically Einige Adverbien bilden unregelmäßige Formen.

Adjective	Adverb	Adverb	
! late	**late** (spät)	**lately** (in letzter Zeit)	Einige Adverbien bilden mehrere Formen.
hard	**hard** (hart)	**hardly** (kaum)	
near	**near** (nah)	**nearly** (beinahe)	

Verwendung

There are many **new computers** on the market. **They** are still **expensive**.	Adjektive beziehen sich auf Nomen oder Pronomen.
We can't **control** the technology **exactly**. Comics are **really popular** with children. Modern computers work **extremely quickly**. **Fortunately, she e-mailed the information to me in time.**	Adverbien beziehen sich auf – Verben – Adjektive – Adverbien – ganze Sätze
! He **felt nervous** about his new job.	Nach Verben, die einen Zustand darstellen (z. B. *feel, be, become, get, seem, appear, keep*) folgt ein Adjektiv.

Word order with adverbs and adverbials (Wortstellung bei Adverbien und adverbialen Bestimmungen)

Die Stellung der Adverbien und adverbialen Bestimmungen (a. B.) ist im Englischen sehr flexibel. Folgende Stellungen sind allerdings üblich:	
At nine o'clock he caught the bus **to work**.	Adverbien/a. B. der Zeit stehen am Anfang oder Ende des Satzes. Adverbien/a. B. des Ortes stehen in der Regel am Ende des Satzes.
Mr Woods **always** travels to work by train. The train arrives **punctually**.	Adverbien/a.B. der Häufigkeit stehen vor dem (Haupt-)Verb. Adverbien/a. B. der Art und Weise stehen in der Regel nach dem Verb (den Verben).
The bus drove **slowly** through **London during the rush hour**.	Treten mehrere Adverbien/a.B. zur gleichen Zeit auf, gilt die Regel: Art und Weise vor Ort und Zeit.

Some or any?

Verwendung

I need **some** help with this program. **Some** newspapers in Britain only appear on Sundays.	in Aussagesätzen vor unbestimmten Mengen (etwas) oder Anzahlen (einige)
They didn't make **any** money with the new campaign. Some TV stations do not show **any** adverts.	in verneinten Sätzen vor unbestimmten Mengen oder Anzahlen (kein oder keine)
Have they done **any** research on the new product? Have you got **any** ideas for a new logo?	in Fragesätzen vor unbestimmten Mengen (etwas) oder Anzahlen (welche?)
Would you like me to show you **some** examples?	in Fragesätzen, wenn sie eine Aufforderung, Bitte oder ein Angebot enthalten
We never needed **any** electronic devices in the past.	nach Wörtern mit negativer Bedeutung

Some/any + -one/-body/-thing/-where

Somebody wants to see you. We haven't been **anywhere** this weekend. Have you got **anything** to say?	Die Regeln zum Gebrauch von *some* und *any* gelten auch für diese Formen.

Much – many – a lot of

Form

much **a lot of**	viel	nicht zählbar
many **a lot of**	viele	zählbar

Verwendung

He spends **a lot of** time practising his communication skills. **A lot of** (many) people have computers in their homes.	In Aussagesätzen wird in der Regel *a lot of* bevorzugt.
I don't have **a lot of/much** money to spend on computers. There aren't **a lot of/many** experts on the staff.	In verneinten Sätzen kann man beide Formen gebrauchen.
How **much** money do you earn? How **many** adverts did they show?	Nach *how* müssen *much* oder *many* stehen.

Little/a little – few/a few

Form

We had **little** help during our training. John had only **a little** time to explain.	wenig ein bisschen	unzählbar
Few computers never break down. There are **a few** good adverts on TV.	wenige ein paar	zählbar

Modal auxiliaries (Modale Hilfsverben)

Form

Modale Form	Ersatzform	Bedeutung
must	**have to**	müssen
can	**be able to**	können
can	**be allowed to**	dürfen
mustn't **can't**	**not be allowed to**	nicht dürfen
can't	**not be able to**	nicht können
needn't	**not have to**	nicht brauchen/nicht müssen

Verwendung

Cars **must/have to** stop at red lights. You **needn't/don't have to** pay in that car park. John **can't/isn't able to** find a parking space.	In der Gegenwart kann entweder die modale oder die Ersatzform benutzt werden.
We **had to** turn right at the traffic lights. He **didn't have to** wait as the lights were green. Jane **couldn't/wasn't allowed to** leave her car outside the house, as there was a "No Parking" sign.	Mit Ausnahme von *could* oder *couldn't* können die modalen Hilfsverben keine anderen Zeiten bilden. Für die anderen Zeiten werden die Ersatzformen verwendet.

Defining relative clauses (Notwendige Relativsätze)

Verwendung

Ein Relativsatz ist notwendig, wenn die darin enthaltene Information nötig ist, um zu erkennen, welche Person oder Sache gemeint ist.	
The man **who/that** is getting on the bus is on his way to the office.	Bei Personen werden die Relativpronomen *who* oder *that* verwendet.
The bus **which/that** is waiting at the traffic lights is going to London.	Bei Sachen wird *which* oder *that* verwendet.
He greeted the man **(who)** he saw on the platform. He read the paper **(which)** he had bought that morning.	Wenn das Relativpronomen Objekt des Satzes ist, kann es weggelassen werden.

Non-defining relative clauses (Nicht notwendige Relativsätze)

Verwendung

Ein Relativsatz ist nicht notwendig, wenn die darin enthaltene Information nicht nötig ist, um zu erkennen, welche Person oder Sache gemeint ist.	
Henry's wife, **who is 58**, was waiting for him at the station.	Ein Relativsatz wird durch Kommas vom Hauptsatz getrennt.
The train, **which was an Intercity**, left punctually.	*that* darf hier nicht verwendet werden.
Henry, **who I met at the station**, is retiring next week.	Wenn das Relativpronomen Objekt eines Relativsatzes ist, darf es nicht weggelassen werden.

The Tenses (Die Zeiten)

Present simple

Form

I/you/we/they he/she/it	**live** **lives**	in Leeds.	Aussage
I/you/we/they he/she/it	**do not (don't) live** **does not (doesn't) live**	in Birmingham.	Verneinung

Do **Does**	I/you/we/they he/she/it	**(not)**	**live**	in Liverpool?	(verneinte) Frage
Where **Why**	**do** you **does** she	**live**? **smoke**?			

!

Who **How many students**	**likes** **live**	basketball? at home?	Ist das Fragewort Subjekt oder Teil des Subjekts, wird *do/does* nicht verwendet.

Besonderheiten bei der Schreibung der s-Endung

go, do watch, kiss carry, try	go**es**, do**es** watch**es**, kiss**es** carr**ies**, tr**ies**

Verwendung

She **has** a brother.	bei einem Dauerzustand
She **(usually) goes** to college by bus. He **(often) helps** at home.	bei regelmäßigen oder wiederholten Handlungen (oft mit Häufigkeitsadverbien wie *usually, normally, often, sometimes, never, always*)
She **doesn't smoke**.	bei Gewohnheiten
He **has** a flat in Mayfield. I **think** that this **is** a good idea.	bei bestimmten Verben, wenn sie einen Zustand beschreiben, z.B. *be, have, look, think, see, know, like, want*
The train **arrives** at 8 o'clock tomorrow morning.	bei Fahrplänen und Veranstaltungsprogrammen (mit Zukunftsbezug)

Present continuous

Form

I you/we/they he/she/it	**am** **are** **is**	**(not)**	**reading** a book.	Aussage und Verneinung

Kurzform der Verneinung: aren't, isn't

Am **Are** **Is**	I you/we/they he/she/it	**(not)**	**reading** a magazine?	Frage und verneinte Frage

Besonderheiten bei der Schreibung

come, take sit, run	com**ing**, tak**ing** sit**ting**, run**ning**

Verwendung

They **are talking** about their courses.	bei Handlungen, die im Moment des Sprechens stattfinden
She **is studying** for her exam this year.	bei Handlungen, die vorübergehend stattfinden, nicht aber unbedingt im Augenblick
She **is having** a good time. A lot of people **are thinking** about their future.	bei bestimmten Verben (siehe *present simple*), wenn sie eine vorübergehende Handlung und keinen Zustand beschreiben
I **am meeting** him tomorrow.	bei zukünftigen Handlungen (mit Zeitbestimmung)
Some students **are always complaining**.	zur gefühlsbetonten Darstellung von Handlungen in Verbindung mit *always*

Past simple

Form

She You	work**ed** need**ed**	in Glasgow a new car.	last year.	Aussagen Grundform + -ed (Infinitiv)

Besonderheiten

mov**e** tr**y** sto**p** go meet	mov**ed** tr**ied** sto**pped** **went** **met**	stummes -e entfällt Konsonant + y wird zu ie Lautverdopplung unregelmäßige Verben haben besondere Formen für das *simple past* dabei wird die 2. Form verwendet

He We	**did not** **(didn't)**	**need** **lose**	special software. the money.	Verneinung

Did	you they it	**(not)**	**move** **live** **take**	to Glasgow? in Birmingham? more than 6 months?	(verneinte) Frage
Why did	he	**(not)**	**stop**	in Liverpool?	

Verwendung

Jane **moved** to Glasgow 7 years ago. First she **worked** for a computer company.	bei Handlungen zu einem bestimmten Zeitpunkt oder in einem abgeschlossenen Zeitraum in der Vergangenheit (oft mit einer Zeitbestimmung wie z. B. *yesterday, last year, in 1988, … ago, from … until*)

Past continuous

Form

I/He/She/It	**was**	**(not)** **(wasn't)**	**standing**	on the platform.	(verneinte) Aussage
We/You/They	**were**	**(not)** **(weren't)**			

Verwendung

As/while he **was driving** to the airport, it started to rain. A lot of passengers **were waiting** at the counter when Jason went into the bank.	für eine Handlung, die schon im Gange war, als eine neue Handlung eintrat
! While I **was reading** a book, my friend **was writing** a letter.	Bei gleichzeitigem Verlauf zweier Handlungen steht die Verlaufsform in beiden Fällen.
! What **did** you **do** when Susan **came** in? When she **came** in I **put** down my newspaper and talked to her.	Für Handlungen, die nacheinander stattfanden, wird in beiden Fällen *past simple* benutzt

Present perfect simple

Form

I/You/We/They	**have**	**(not)** **(haven't)**	**had** a car for three years.	(verneinte) Aussage *have/has* + 3. Form
He/She/It	**has**	**(not)** **(hasn't)**	**sold** 20 million CDs. **been** to Scotland.	

Have	I/you/we/they	**(not)**	**made** a loss? **taken** a photo? **been** here?	(verneinte) Frage
Has	he/she/it			

Verwendung

I **have repaired** the car.	bei Handlungen in der Vergangenheit ohne Zeitangabe – das Ergebnis ist oft wichtiger als die Zeitangabe
They **have sold** 50,000 cars **up till now**.	mit Zeitbestimmungen, die einen Zeitraum beschreiben, der noch andauert, z. B. *today, this week, so far, in the last ten years*
He has **already paid** the bill.	mit bestimmten Adverbien wie z. B. *ever, never, always, yet, just, already*
How long have you **known** about this book? The Browns **have lived** there **for** 3 years. We **have lived** here **since** 1986.	bei nicht abgeschlossenen Zuständen mit *how long, since* und *for* – *since* bezieht sich dabei auf einen Zeitpunkt, *for* auf einen Zeitraum

Present perfect continuous

Form

I/you/we/they	**have**	**been standing.**	Present simple von *have* + *been* + -ing-Form

Verwendung

How long have you been waiting? I've been waiting – for half an hour. – since ten o'clock.	bei Handlungen, die in der Vergangenheit begannen und noch andauern – meist in Verbindung mit *how long, since, for*

Past perfect simple

Form

I/You/He/She/ It/We/They	**had**	**(not)** **(hadn't)**	**travelled** to Paris.	(verneinte) Aussage

Verwendung

After he **had arrived** at the hotel, he went to his room. When I arrived at the station, the ticket office **had** already **closed**.	für Handlungen oder Zustände, die vor einem Zeitpunkt in der Vergangenheit abgeschlossen waren

Future with 'going to'

Form

I He/She/It We/You/They	**am ('m)** **is ('s)** **are ('re)**	**(not)**	**going to** watch TV.	(verneinte) Aussage

Am **Is** **Are**	I he/she/it we/you/they	**going to have**	a cup of coffee?	Frage

Verwendung

She **is going to answer** the phone.	bei Absichten und Vorhaben
That fax machine **is going to break** down.	bei Vorhersagen, die aufgrund bereits bekannter Fakten oder bisheriger Erfahrungen sicher oder logischerweise in Erfüllung gehen müssen

Future with 'will'

Form

I/You He/She/It We/They	**will** **will not (won't)**	**use** alternative energy. **waste** water.	Aussage Verneinung

Will	I/you he/she/it we/they	**use** talking computers?	Frage

Verwendung

Computers **will probably do** most of the work.	bei Vorhersagen mit /*suppose/expect* und *probably*
Just a minute, **I'll help** you with your shopping.	bei Entscheidungen, die im Moment des Sprechens getroffen werden
I'll phone you tomorrow, I promise.	bei Versprechen
Will you **make** some coffee, please.	um eine Bitte auszudrücken

Question tags (Frageanhängsel)

Form

Aussagesatz	Frageanhängsel	
You can drive,	**can't you?**	Das Hilfsverb wird in dem Frageanhängsel wiederholt. Ist der Aussagesatz positiv, so wird das Frageanhängsel verneint.
Pollution is increasing,	**isn't it?**	
Trams will improve our public transport service,	**won't they?**	
Statistics aren't always correct,	**are they?**	Ist der Aussagesatz verneint, so wird das Frageanhängsel bejaht.
You mustn't drive over 70 mph on motorways,	**must you?**	
We enjoyed the journey,	**didn't we?**	Wenn der Aussagesatz kein Hilfsverb enthält, wird das Frageanhängsel mit *to do* in der Zeit des Aussagesatzes gebildet.
She usually goes to Britain,	**doesn't she?**	

Verwendung

	Frageanhängsel werden verwendet:
Air travel is really comfortable, **isn't it?**	beim Wunsch nach Bestätigung
You didn't walk here, **did you?**	bei Überraschung
Oh, it's not another traffic jam, **is it?**	bei Verärgerung
Well, that's a really new idea, **isn't it?**	bei ironischen Aussagen

Reported speech (Indirekte Rede)

Form

Aussagen

	Die Zeitformen ändern sich:
"We are protesting about pollution." Harry said they **were protesting** about pollution. "He doesn't buy cans." June told me he **didn't buy** cans.	*Present tenses* werden zu *past tenses*.
"Watsons opened four supermarkets in 1995." Fred pointed out that Watsons **had opened** four supermarkets in 1995. "They have started a bus service." He added that they **had started** a bus service.	*Simple past* und *present perfect* werden zu *past perfect*.
"There will be more 'green' products." Sue went on to say that there **would be** more 'green' products.	*will* wird zu *would*.
"I can save more energy." Harry told me that **he** could save more energy.	Die Pronomen werden angepasst.
"We are meeting the town planner tomorrow." Sue added that they were meeting the town planner **the next day**. "I bought the car three days ago." He said that he had bought the car **three days before**.	Orts- (z. B. *here* wird zu *there*) und Zeitbestimmungen werden angepasst.

Fragen

"What are you doing about waste?" **They asked what we were doing** about waste. "When did you introduce bicycle routes?" **We wanted to know when they had introduced** bicycle routes.	Bei Fragen wird die Wortstellung geändert.

Verwendung

Indirekte Rede wird benutzt, wenn berichtet werden soll, was jemand gesagt oder gefragt hat.

Conditionals (Bedingungssätze)

Form

Type 1

If-Satz	Hauptsatz	
If **we fit** a shower,	**we will** save water.	*Present simple* wird für das Verb im If-Satz, Futur mit *will* für das Verb im Hauptsatz verwendet.

Type 2

If-Satz	Hauptsatz	
If the government **developed** solar energy,	there **would be** less pollution.	*Past simple* wird für das Verb im If-Satz, *Conditional I* für das Verb im Hauptsatz verwendet.
If I **were** Environment Minister,	I **could** put my ideas about energy into practice.	*could* oder *might* sind Alternativen zu *would* im Hauptsatz.

Type 3

If-Satz	Hauptsatz	
If Tom and Kate **had insulated** their roof years ago,	they **would have saved** a lot of money.	*Past perfect* wird für das Verb im If-Satz, *Conditional II* für das Verb im Hauptsatz verwendet.

Verwendung

Type 1 wird gebraucht, wenn die Voraussetzungen für die Erfüllung der Bedingung schon existieren und es deswegen wahrscheinlich ist, dass die Handlung im If-Satz in Erfüllung gehen wird (wir planen, eine Dusche einzubauen).
Type 2 wird gebraucht, wenn a) wir nicht erwarten, dass die Handlung im If-Satz eintreten wird. Wir stellen uns die Situation rein theoretisch vor. b) die Handlung im If-Satz gar nicht eintreten kann, weil sie in Gegensatz zu den Tatsachen steht (ich bin nicht der Umweltminister).
Type 3 wird gebraucht, wenn die Handlung im If-Satz nicht eintreten kann, weil sie in der Vergangenheit liegt. Es kann an der Situation nichts mehr geändert werden.

Passive voice (Passiv)

Form

The robots **are equipped** with sensors. This motor **was not made** in Great Britain. **Will** the new car **be sold** in Japan?	Form von *be* in der jeweiligen Zeit + Partizip Perfekt (3. Form) des Verbs
Now the wheels **can be fitted**. The car **must be tested** first.	auch mit Hilfsverben

Verwendung

These cars **were made** in Germany. The handbook **is written** in English.	hauptsächlich in der Schriftsprache, wenn der Ausführende unbekannt, unwichtig oder selbstverständlich ist
This information **is checked by** the central computer.	Wenn der Ausführende allerdings genannt werden soll, dann benutzt man die Präposition *by* *(by-agent)*.
It is said **It is reported** **It is believed** **It is supposed** that computers will become even more important in the future. **He is said** to be rich.	bei Verben des Meinens und Berichtens
English **is spoken** here. = Man spricht Englisch. We **will be told** the result later. = Man wird uns später das Ergebnis sagen.	im Deutschen wird ein Aktivsatz mit ‚man' bevorzugt

Infinitive (Infinitiv)

Verwendung

ohne *to*

We **must write** a report about our trip to England.	nach den meisten Hilfsverben
Our new teachers **make us work** very hard, but on Friday they **let us go** home early.	nach *make* und *let* + direktem Objekt

mit *to*

The teacher **asked me to show** my photos of Newcastle. She **told them to wait** outside for a moment.	nach bestimmten Verben (*ask/tell/advise/expect* etc.)
This is not **easy to understand**. I am **surprised to hear** that.	nach Adjektiven
I do not know **what to do**. I soon found out **where to go**.	nach Fragewörtern

Gerund (Gerundium)

Verwendung

Drinking and **driving** is dangerous.	als Subjekt
I **enjoy swimming.**	als Objekt nach Verben ohne Präpositionen, wenn eine allgemeingültige Situation beschrieben wird: z. B. *enjoy/like/dislike/hate/stop/start/avoid/ suggest/mind/love/recommend/prefer*
I **look forward to seeing** you again. He is **tired of waiting** here. They saw the **danger of destroying** the environment.	nach Verben/Adjektiven/Substantiven mit Präpositionen
He has helped me a lot **by giving** me that map of Newcastle. The child crossed the road **without looking**.	nach Präpositionen mit adverbialer Bedeutung, z. B. *by/without/instead of*
! Normally I enjoy swimming but today I **would prefer to play** tennis. I remember learning those rules. = Ich erinnere mich daran, dass ich die Regeln gelernt habe. I must remember to learn these rules. = Ich muss daran denken, diese Regeln zu lernen.	aber kein Gerundium, wenn es sich um eine Ausnahmesituation handelt (oft mit *would*) oder wenn die Bedeutung es nicht zulässt, z. B. bei *stop/start/remember*

Participle (Partizip)

Form

Das Partizip besteht aus zwei Formen

waiting/going/watching	Partizip Präsens
waited/gone/watched	Partizip Perfekt

Verwendung

The weather forecast for the **coming** week … The weather in New York last week was rather **mixed**.	als Adjektiv
You can find several families **living** in one flat. There are a lot of restaurants there especially **built** for the tourists.	anstatt von Relativsätzen
After talking to the press they started their demonstration. **Having discussed** the problem with the President personally, they flew home to California.	als Verkürzung von Adverbialsätzen (*after/before/because/while* etc.)

Unit vocabulary

AE = American English
BE = British English

Unit 1

Starter

comprehensive school [ˌkɒmprɪˈhensɪvˌskuːl]	Gesamtschule
GCSE [ˌdʒiːsiːesˈiː] = General Certificate of Secondary Education [ˌdʒenərəl səˈtɪfɪkət əv ˌsekəndrɪ‿ˌed ʒʊˈkeɪʃn]	etwa: Fachoberschulreife
certificate [səˈtɪfɪkət]	Zeugnis
studies [ˈstʌdɪz]	Kurs, Studium
engineering studies [ˌendʒɪˈnɪərɪŋ ˌstʌdɪz]	Ingenieurwissenschaft
social studies [ˈsəʊʃl ˌstʌdɪz]	Sozialwissenschaften
business studies [ˈbɪznɪs‿ˌstʌdɪz]	etwa: Fachbereich Wirtschaft
BTEC National Diploma [ˌbiːtiːiːˈsiː ˌnæʃənl dɪˈpləʊmə]	etwa: Fachhochschulreifeprüfung
public service [ˌpʌblɪk ˈsɜːvɪs]	öffentlicher Dienst

A1

full-time [ˌfʊl ˈtaɪm]	Vollzeit
part-time [ˌpɑːt‿ˈtaɪm]	Teilzeit
to run courses [rʌn ˈkɔːsɪz]	Kurse anbieten
qualification [ˌkwɒlɪfɪˈkeɪʃn]	Qualifikation, Befähigung
level [ˈlevl]	Stufe, Niveau
bus pass [ˈbʌs ˌpɑːs]	Busausweis
grant [grɑːnt]	Zuschuss, Bafög
equipment [ɪˈkwɪpmənt]	Ausrüstung, Ausstattung
electrical engineering [ɪˈlektrɪkl‿ˌendʒɪˈnɪərɪŋ]	Elektrotechnik
mechanical engineering [məˈkænɪkl‿ˌendʒɪˈnɪərɪŋ]	Maschinenbau
civil engineering [ˈsɪvl‿ˌendʒɪˈnɪərɪŋ]	Bauwesen
general science [ˌdʒenərəl ˈsaɪəns]	allgemeine Naturwissenschaften
application [ˌæplɪˈkeɪʃn]	Anwendung; Bewerbung
CAD [ˌsiːeɪˈdiː] (Computer Aided Design) [kəmˈpjuːtər‿ˌeɪdɪd‿dɪˈzaɪn]	Computerunterstützte Zeichnung
CAM [ˌsiːeɪˈem] (Computer Aided Manufacturing) [kəmˈpjuːtər‿ˌeɪdɪd ˌmænjəˈfæktʃərɪŋ]	Computerunterstützte Fertigung
industrial [ɪnˈdʌstrɪəl]	industriell
control [kənˈtrəʊl]	Kontrolle, hier: Regeltechnik
instrumentation [ˌɪnstrəmenˈteɪʃn]	Messtechnik
office skills [ˈɒfɪs‿ˌskɪlz]	Bürofertigkeiten
accountancy [əˈkaʊntənsɪ]	Rechnungswesen
statistics [stəˈtɪstɪks]	Statistik
finance [ˈfaɪnæns]	Finanzwesen
sociology [ˌsəʊʃɪˈɒlədʒɪ]	Soziologie
psychology [saɪˈkɒlədʒɪ]	Psychologie
principle [ˈprɪnsəpl]	Grundsatz, Prinzip
health care [ˈhelθ ˌkeə]	Gesundheitsfürsorge
paragraph [ˈpærəgrɑːf]	Absatz

A2

campus [ˈkæmpəs]	Campus, Universitätsgelände
long time no see [lɒŋ ˌtaɪm nəʊ ˈsiː]	lange nicht gesehen
formula [ˈfɔːmjələ]	Formel
packing [ˈpækɪŋ]	Verpackung
pocket money [ˈpɒkɪt ˌmʌnɪ]	Taschengeld
beginner [bɪˈgɪnə]	Anfänger(in)
to rush [rʌʃ]	eilen
information sheet [ˌɪnfəˈmeɪʃn ˌʃiːt]	Informationsblatt
female [ˈfiːmeɪl]	weiblich
male [meɪl]	männlich
dialogue [ˈdaɪəlɒg]	Dialog
bedsitter [bedˈsɪtə]	Einzimmerwohnung
coach [kəʊtʃ]	Trainer

C1

exchange [ɪksˈtʃeɪndʒ]	Austausch
host [həʊst]	Gastgeber(in)
adventure [ədˈventʃə]	Abenteuer
head of department [ˌhed‿əv dɪˈpɑːtmənt]	hier: Fachbereichsleiter(in)

to get to know [get‿tʊ 'nəʊ]	kennenlernen
to put (s.o./sth.) to the test [ˌpʊt‿tʊ ðə 'test]	jmdn./etwas auf die Probe stellen
to be into [biː‿'ɪntuː]	Fan von etwas sein
smoker ['sməʊkə]	Raucher(in)
metal ['metl]	Metall
heavy metal [ˌhevɪ 'metl]	Heavy Metal (harte Rockmusik)
to remark [rɪ'mɑːk]	bemerken
biscuit ['bɪskɪt]	Keks
as well as [əz‿'wel‿əz]	sowie
to cope (with) [kəʊp]	mit etwas klar kommen
to get on like a house on fire [get‿'ɒn laɪk‿əˌhaʊs‿ɒn 'faɪə]	sich sehr gut verstehen
moor [mɔː]	Moor
district ['dɪstrɪkt]	Gegend
joint [dʒɔɪnt]	gemeinsam
project ['prɒdʒekt]	Projekt
workshop ['wɜːkʃɒp]	Werkstatt, Arbeitsgruppen
organizer ['ɔːgənaɪzə]	Veranstalter
staffroom ['stɑːfruːm]	Lehrerzimmer
comment ['kɒment]	Bemerkung, Kommentar
to broaden the mind [ˌbrɔːdn ðə 'maɪnd]	den Horizont erweitern
sights [saɪts]	Sehenswürdigkeiten
bold type [ˌbəʊld‿'taɪp]	Fettdruck, fett gedruckt

C2

informal [ɪn'fɔːml]	informell
scheme [skiːm]	Projekt, Plan
Yours [jɔːz]	Liebe Grüße (Brief)
personality [ˌpɜːsə'nælətɪ]	Persönlichkeit
to look like ['lʊk laɪk]	aussehen wie
can't stand [kɑːnt‿'stænd]	nicht ausstehen können

D1

enthusiastic [ɪnˌθjuːzɪ'æstɪk]	begeistert
hard-working [ˌhɑːd'wɜːkɪŋ]	fleißig
phrase [freɪz]	Ausdruck
locals ['ləʊklz]	Einheimische
a bit [ə bɪt]	ein bisschen
in my view [ɪn‿'maɪ ˌvjuː]	meiner Ansicht nach
for instance [fər‿'ɪnstəns]	beispielsweise

Tapescript

electronics factory [ˌelek'trɒnɪks ˌfæktrɪ]	Elektronikfabrik
power station ['paʊə ˌsteɪʃn]	Elektrizitätswerk, Kraftwerk

Unit 2

Starter

to discover [dɪ'skʌvə]	entdecken
to work out [wɜːk‿'aʊt]	herausfinden, ausarbeiten, lösen
suitable ['suːtəbl]	geeignet
on your own [ˌɒn jər‿'əʊn]	allein
per [pə]	pro
mark [mɑːk]	Zensur, Note
novel ['nɒvl]	Roman
puzzle ['pʌzl]	Rätsel
active ['æktɪv]	aktiv
practical ['præktɪkl]	praktisch
sportsman ['spɔːtsmən]	Sportler
sportswoman ['spɔːtsˌwʊmən]	Sportlerin
officer ['ɒfɪsə]	Beamte(r), Offizier(in)
creative [krɪ'eɪtɪv]	kreativ
singer ['sɪŋə]	Sänger(in)
dancer ['dɑːnsə]	Tänzer(in)
hairdresser ['heəˌdresə]	Friseur(in)
flight attendant ['flaɪt‿əˌtendənt]	Flugbegleiter(in)
office worker ['ɒfɪs ˌwɜːkə]	Büroangestellte(r)
tourist guide ['tʊərɪst ˌgaɪd]	Fremdenführer(in)
salesman/saleswoman ['seɪlzmən /'seɪlzˌwʊmən]	Verkäufer(in)
independent [ˌɪndɪ'pendənt]	unabhängig
logical ['lɒdʒɪkl]	logisch
well-paid [ˌwel 'peɪd]	gut bezahlt
specialist ['speʃəlɪst]	Spezialist(in)
advertising manager ['ædvətaɪzɪŋ ˌmænɪdʒə]	Werbeleiter(in)

A1

unlimited company [ʌnˌlɪmɪtɪd 'kʌmpənɪ]	Gesellschaft mit unbeschränkter Haftung
alive [ə'laɪv]	lebend, lebendig, aktiv

sales [seɪlz]	Verkaufszahlen
enormous [ɪ'nɔːməs]	gewaltig, enorm
turnover ['tɜːnˌəʊvə]	Umsatz
to graduate ['grædjʊeɪt]	Schulabschluss machen, Schule oder Universität beenden, einen akad. Grad erwerben
degree [dɪ'griː]	akad. Grad (z. B. B.A., M.A.)
self-employed [ˌselfɪm'plɔɪd]	selbstständig
risk [rɪsk]	Risiko
moreover [mɔː'rəʊvə]	überdies, außerdem, zudem, ferner
wage [weɪdʒ]	Lohn
to service ['sɜːvɪs]	(Auto, Maschine etc.) warten
to expand [ɪk'spænd]	ausdehnen, ausbreiten
in the meantime [ɪn ðə 'miːntaɪm]	in der Zwischenzeit
natural ['nætʃərəl]	natürlich
to sum up [sʌm 'ʌp]	zusammenfassen
to adapt [ə'dæpt]	bearbeiten
noun [naʊn]	Substantiv
adjective ['ædʒɪktɪv]	Adjektiv
to a great extent [tʊ‿ə ˌgreɪt‿ɪk'stent]	in großem Umfang
fortune ['fɔːtʃuːn]	Glück, Reichtum

A2

video production ['vɪdɪəʊ prəˌdʌkʃn]	Videoproduktion
listener ['lɪsnə]	Zuhörer(in)
to start up (business) [ˌstɑːt ʌp 'bɪznɪs]	(Geschäft) anfangen
thought [θɔːt]	Gedanke
to make up one's mind [ˌmeɪk ʌp wʌnz 'maɪnd]	sich entschließen
to record [rɪ'kɔːd]	aufnehmen (Ton, Video etc.)
wedding ['wedɪŋ]	Hochzeit
income ['ɪŋkʌm]	Einkommen
to close down [kləʊz 'daʊn]	schließen, zumachen (z.B. Firma)
main [meɪn]	Haupt-
market development [ˌmɑːkɪt dɪ'veləpmənt]	Marktentwicklung
photographer [fə'tɒgrəfə]	Fotograf(in)
to afford [ə'fɔːd]	sich leisten
tense [tens]	Zeitform
to link; link [lɪŋk]	verbinden; Verbindung

C1

to have in common [hæv ɪn 'kɒmən]	gemeinsam haben
Bank Holiday [ˌbæŋk 'hɒlɪdeɪ]	(öffentlicher) Feiertag
working class [ˌwɜːkɪŋ 'klɑːs]	Arbeiterklasse
condition [kən'dɪʃn]	Bedingung
except [ɪk'sept]	außer
underpaid [ˌʌndə'peɪd]	unterbezahlt
survey ['sɜːveɪ]	Untersuchung, Umfrage
employee [ɪm'plɔɪiː]	Arbeitnehmer(in), Angestellte(r)
attitude ['ætɪtjuːd]	Haltung, Einstellung
workmate ['wɜːkmeɪt]	Arbeitskollege, -kollegin
benefit ['benɪfɪt]	Zuwendung, soziale Leistung
in general [ɪn 'dʒenərəl]	im Allgemeinen
to get on with [get 'ɒn wɪð]	auskommen mit
to identify [aɪ'dentɪfaɪ]	identifizieren
satisfied ['sætɪsfaɪd]	zufrieden
Dutch [dʌtʃ]	niederländisch; Niederländer(in)
sick leave ['sɪk ˌliːv]	krankheitsbedingte Abwesenheit
working hours ['wɜːkɪŋ‿ˌaʊəz]	Arbeitszeit
to vary ['veərɪ]	sich ändern, unterschiedlich sein
amount [ə'maʊnt]	Betrag, Menge
average ['ævərɪdʒ]	Durchschnitt
satisfaction [ˌsætɪs'fækʃn]	Zufriedenheit
to carry out [ˌkærɪ‿'aʊt]	ausführen

C2

would like [wʊd 'laɪk]	würde(n) gerne

D

step [step]	Schritt, Stufe
leading ['liːdɪŋ]	führend
to require [rɪ'kwaɪə]	benötigen, erfordern
responsible [rɪ'spɒnsəbl]	verantwortlich
assistant [ə'sɪstənt]	Assistent(in)
export assistant ['ekspɔːt‿əˌsɪstənt]	etwa: Außenhandelskaufmann, -frau
childcare assistant ['tʃaɪldkeər‿əˌsɪstənt]	Kinderpfleger(in)
candidate ['kændɪdət]	Kandidat(in), Bewerber(in)
aged … [eɪdʒd]	… Jahre alt

similar ['sɪmələ]	ähnlich
essential [ɪ'senʃl]	wesentlich, unentbehrlich
willing ['wɪlɪŋ]	bereit
overseas [ˌəʊvə'siːz]	in Übersee
to care for [keə fɔː]	sich um etw. kümmern
salary ['sælərɪ]	Gehalt
applicant ['æplɪkənt]	Bewerber(in)
CV [ˌsiː'viː] = curriculum vitae [kəˌrɪkjələm 'viːtaɪ]	Lebenslauf
to apply [ə'plaɪ]	sich bewerben
Dear Sir [dɪə 'sɜː]	Sehr geehrter Herr
Dear Madam [dɪə 'mædəm]	Sehr geehrte Dame
with reference to [wɪð 'refrəns tuː]	mit Bezug auf
reference ['refrəns]	Referenz, Hinweis
vacancy ['veɪkənsɪ]	freie Stelle
nationality [ˌnæʃə'nælətɪ]	Nationalität
task [tɑːsk]	Aufgabe
to enclose [ɪn'kləʊz]	beilegen (als Anlage)
to look forward to (-ing) [lʊk 'fɔːwəd‿tuː]	sich freuen auf (Schlussformel im Geschäftsbrief)
Yours faithfully [jɔːz 'feɪθfʊlɪ]	Hochachtungsvoll
Yours sincerely [jɔːz‿sɪn'sɪəlɪ]	Mit freundlichen Grüßen

Tapescript

challenge ['tʃæləndʒ]	Herausforderung
to involve [ɪn'vɒlv]	einschließen

Unit 3

Starter

European Commission [ˌjʊərəpɪən kə'mɪʃn]	Europäische Kommission
citizen ['sɪtɪzn]	Bürger(in)
capital city [ˌkæpɪtl 'sɪtɪ]	Hauptstadt
European Union [ˌjʊərəpɪən‿'juːnjən]	Europäische Union (die EC wurde im November 1993 in die EU umbenannt)

A1

Warsaw ['wɔːsɔː]	Warschau
Poland ['pəʊlənd]	Polen
Polish ['pəʊlɪʃ]	polnisch
seminar ['semɪnɑː]	Seminar
lecture ['lektʃə]	Vorlesung
youth hostel ['juːθ ˌhɒstl]	Jugendherberge
bureaucracy [bjʊ'rɒkrəsɪ]	Bürokratie
to fill in [fɪl 'ɪn]	ausfüllen
to insure [ɪn'ʃʊə]	versichern
discount ['dɪskaʊnt]	Rabatt
entrance exam ['entrəns‿ɪgˌzæm]	Aufnahmeprüfung
economics [ˌiːkə'nɒmɪks]	Wirtschaft(swissenschaft), Volkswirtschaft, Volkswirtschaftslehre
to know by name [ˌnəʊ baɪ 'neɪm]	mit Namen kennen
to take a (language) test [teɪk ə 'test]	einen Test machen, an einer (Sprach-)Prüfung teilnehmen
to reserve [rɪ'zɜːv]	reservieren

A2

unity ['juːnətɪ]	Einheit
process ['prəʊses]	Vorgang, Prozess
democratic [ˌdemə'krætɪk]	demokratisch
peaceful ['piːsfʊl]	friedlich
wealthy ['welθɪ]	wohlhabend
political [pə'lɪtɪkl]	politisch
factor ['fæktə]	Faktor
to slow down [sləʊ 'daʊn]	verlangsamen
column ['kɒləm]	Spalte
past tense [ˌpɑːst‿'tens]	Vergangenheitsform
Belgium ['beldʒəm]	Belgien
Germany ['dʒɜːmənɪ]	Deutschland
Italy ['ɪtəlɪ]	Italien
Luxembourg ['lʌksəmbɜːg]	Luxemburg
(the) Netherlands [ðə 'neðələndz]	Niederlande
Denmark ['denmɑːk]	Dänemark
Ireland ['aɪələnd]	Irland
Great Britain [ˌgreɪt‿'brɪtn]	Großbritannien
election [ɪ'lekʃn]	Wahl
parliament ['pɑːləmənt]	Parlament
Greece [griːs]	Griechenland
Spain [speɪn]	Spanien
state [steɪt]	Staat
community [kə'mjuːnətɪ]	Gemeinschaft, Bevölkerungsgruppe, Gemeinde
treaty ['triːtɪ]	Vertrag
to come into operation ['kʌm ˌɪntʊ‿ˌɒpə'reɪʃn]	in Kraft treten
single ['sɪŋgl]	einzig, hier: einheitlich
currency ['kʌrənsɪ]	Währung

C1

cosmopolitan [ˌkɒsmə'pɒlɪtn]	kosmopolitisch
glass-walled [ˌglɑːs'wɔːld]	mit Wänden aus Glas
Prague [prɑːg]	Prag
neglect [nɪ'glekt]	Vernachlässigung
communist era ['kɒmjənɪst ˌɪərə]	kommunistisches Zeitalter
strength [streŋkθ]	Stärke
Estonia [es'təʊnɪə]	Estland
Baltic ['bɔːltɪk]	baltisch; Ostsee
trading centre ['treɪdɪŋ ˌsentə]	Handelszentrum
run-down [ˌrʌn'daʊn]	heruntergekommen
diversity [daɪ'vɜːsətɪ]	Vielfalt, Verschiedenheit
mayor [meə]	Bürgermeister(in)
to turn into [tɜːn 'ɪntuː]	verwandeln, umwandeln in
huge [hjuːdʒ]	riesig
sunchair ['sʌnˌtʃeə]	Liegestuhl
Millennium Wheel [mɪˌlenɪəm 'wiːl]	Millennium-Riesenrad
millennium [mɪ'lenɪəm]	Jahrtausend
permanent ['pɜːmənənt]	ständig, permanent, dauerhaft
high-speed [ˌhaɪ'spiːd]	Hochgeschwindigkeits-
low-cost [ˌləʊ'kɒst]	preiswert
countryside ['kʌntrɪsaɪd]	Land, Landschaft
leadership ['liːdəʃɪp]	Führung, Leitung
disappearance [ˌdɪsə'pɪərəns]	Verschwinden
boundary ['baʊndərɪ]	Grenze
gradually ['grædʒʊəlɪ]	allmählich
to realise/realize ['rɪəlaɪz]	feststellen, erkennen, bemerken
view [vjuː]	Blick, Sicht, Ansicht, Aussicht
individually [ˌɪndɪ'vɪdʒʊəlɪ]	einzeln

C2

United Kingdom [juːˌnaɪtɪd 'kɪŋdəm]	Vereinigtes Königreich
holidaymaking ['hɒlɪdeɪˌmeɪkɪŋ]	in den Urlaub fahren
rate [reɪt]	Quote, Prozentsatz
domestic trip [dəˌmestɪk 'trɪp]	Inlandsausflug
outbound ['aʊtbaʊnd]	ins Ausland
holidaymaker ['hɒlɪdeɪˌmeɪkə]	Urlauber(in)
least [liːst]	am wenigsten
above [ə'bʌv]	oben

D1

gap year ['gæp ˌjɪə]	etwa: Lückenjahr
to advise [əd'vaɪz]	(be)raten
internship ['ɪntɜːnʃɪp]	Praktikum
exchange year [ɪks'tʃeɪndʒ‿ˌjɪə]	Austauschjahr

D2

to consist [kən'sɪst] of	bestehen aus
to be committed to doing sth. [bɪ kəˌmɪtɪd‿tə 'duːɪŋ ˌsʌmθɪŋ]	sich dafür einsetzen, etw. zu tun
prosperity [prɒs'perɪtɪ]	Wohlstand
intention [ɪn'tenʃn]	Absicht
unique [juː'niːk]	einzigartig
to establish [ɪ'stæblɪʃ]	einrichten, gründen
common ['kɒmən]	gemeinsam
institution [ˌɪnstɪ'tjuːʃn]	Einrichtung
matter ['mætə]	Angelegenheit

Tapescript

competitor [kəm'petɪtə]	Wettbewerber(in)
to file [faɪl]	abheften, ablegen
to gather ['gæðə]	sammeln
to participate [pɑː'tɪsɪpeɪt]	teilnehmen
travel agency ['trævl ˌeɪdʒənsɪ]	Reisebüro

Unit 4

Starter

device [dɪ'vaɪs]	Gerät
communications [kəˌmjuːnɪ'keɪʃnz]	Telekommunikation, Kommunikationswege

A1

thumb [θʌm]	Daumen
text message ['tekst ˌmesɪdʒ]	SMS-Nachricht
junkie ['dʒʌŋkɪ]	Abhängige(r)
addiction [ə'dɪkʃn]	Sucht, Abhängigkeit
to dominate ['dɒmɪneɪt]	dominieren, beherrschen
thankfully ['θæŋkfʊlɪ]	glücklicherweise
mobile phone operator [ˌməʊbaɪl 'fəʊn ˌɒpəreɪtə]	Mobilfunkbetreiber

to promote [prə'məʊt]	fördern, werben für
broadcaster ['brɔːdkɑːstə]	Rundfunksender
to launch [lɔːntʃ]	etw. starten, auf den Markt bringen, einführen
event [ɪ'vent]	Veranstaltungen
promotion [prə'məʊʃn]	(Verkaufs-)Förderung, Promotion, Beförderung
to text [tekst]	simsen, eine SMS-Nachricht senden
obsession [əb'seʃn]	Besessenheit
to affect [ə'fekt]	beeinflussen, sich auswirken auf
shorthand ['ʃɔːthænd]	Kurzschrift
spelling ['spelɪŋ]	Schreibweise, Rechtschreibung
construction [kən'strʌkʃn]	Bau, Bauweise, Konstruktion
to suffer ['sʌfə]	leiden
to claim [kleɪm]	behaupten, verlangen, beanspruchen, Anspruch geltend machen
to get across [get ə'krɒs]	verständlich machen
proper ['prɒpə]	korrekt, angemessen
essay ['eseɪ]	Abhandlung, Aufsatz
to mark down [ˌmɑːk 'daʊn]	(Preise, Noten) herabsetzen

A2

bar graph ['bɑːˌgrɑːf]	Säulendiagramm
graph [grɑːf]	Diagramm, Schaubild, mathematische Kurve
percentage [pə'sentɪdʒ]	prozentualer Anteil
household ['haʊshəʊld]	Haushalt
access ['ækses]	Zugang
broadband ['brɔːdbænd]	Breitband
mobile phone [ˌməʊbaɪl 'fəʊn]	Handy
figure ['fɪgə]	Zahl, Figur

C

share [ʃeə]	Anteil
to recognize ['rekəgnaɪz]	(wieder)erkennen, anerkennen
according to [ə'kɔːdɪŋ tuː]	(seiner/ihrer) Meinung nach, laut
best known [ˌbest'nəʊn]	bekannteste(r, s)
logo ['ləʊgəʊ]	Logo, (Firmen)Emblem
commercial [kə'mɜːʃl]	Werbespot
airtime ['eətaɪm]	Sendezeit
pleased [pliːzd]	zufrieden, erfreut
background ['bækgraʊnd]	Hintergrund; Herkunft, Verhältnisse
tennis serve ['tenɪs ˌsɜːv]	Aufschlag (Tennis)
viewer ['vjʊə]	Zuschauer(in)
remote control [rɪˌməʊt kən'trəʊl]	Fernbedienung
to switch [swɪtʃ]	umschalten, schalten
habit ['hæbɪt]	Gewohnheit
campaign [kæm'peɪn]	Kampagne, Werbefeldzug
advertiser ['ædvəˌtaɪzə]	Inserent(in), Auftraggeber(in) von Werbesendungen
to overestimate [ˌəʊvər'estɪmeɪt]	überschätzen
value ['væljuː]	Wert
to decrease [dɪ'kriːs]	abnehmen, vermindern
worldwide [ˌwɜːld'waɪd]	weltweit
viewpoint ['vjuːpɔɪnt]	Standpunkt
on the one hand [ɒn ðə 'wʌn ˌhænd]	auf der einen Seite
whereas [weə'ræz]	während

D1

superfast [ˌsuːpə'fɑːst]	sehr schnell
superb [suː'pɜːb]	erstklassig, großartig
polite [pə'laɪt]	höflich

D2

advert ['ædvɜːt]	Kurzform von "advertisement" = Werbeanzeige
slogan ['sləʊgn]	Slogan, Werbespruch
brand name ['brænd ˌneɪm]	Markenname
lifestyle ['laɪfstaɪl]	Lebensstil
to associate [ə'səʊʃɪeɪt]	assoziieren, in Verbindung bringen
target group ['tɑːgɪt ˌgruːp]	Zielgruppe
to aim (at) [eɪm]	zielen (auf)

D3

to refer to [rɪ'fɜː tuː]	sich beziehen auf

Tapescript

confident ['kɒnfɪdənt]	hier: sicher
family-sized [ˌfæmlɪ'saɪzd]	für Familien zugeschnitten
local store [ˌləʊkl 'stɔː]	Geschäft vor Ort
marvellous ['mɑːvələs]	fantastisch
mate [meɪt]	Kollege, Kumpel
to stock [stɒk]	auf Lager haben

Unit 5

Starter

to reduce [rɪdjuːs] verringern, senken

A1

congested [kən'dʒestɪd] überfüllt, verstopft
day by day [ˌdeɪ baɪ 'deɪ] Tag für Tag
congestion [kən'dʒestʃn] Stau
motorway ['məʊtəweɪ] Autobahn
major ['meɪdʒə] bedeutend
to get used to [get 'juːst‿tuː] sich gewöhnen an
to look into [lʊk 'ɪntuː] untersuchen, prüfen
technological [ˌteknə'lɒdʒɪkl] technologisch, technisch
innovation [ˌɪnə'veɪʃn] Neuerung
navigation system [ˌnævɪ'geɪʃn ˌsɪstəm] Navigationssystem
tram [træm] Straßenbahn
overcrowded [ˌəʊvə'kraʊdɪd] überfüllt
rush-hour ['rʌʃˌaʊə] Hauptverkehrszeit
traffic jam ['træfɪk ˌdʒæm] Verkehrsstau
fumes [fjuːmz] Rauch, Dämpfe, Abgase

A2

to worry ['wʌrɪ] sich sorgen, sich beunruhigen
audio ['ɔːdɪəʊ] audio, Hör-
touch [tʌtʃ] Berührung
button ['bʌtn] Knopf
to recalculate [ˌriː'kælkjəleɪt] neu berechnen
to guide [gaɪd] leiten, führen
destination [ˌdestɪ'neɪʃn] Reiseziel, Zielort
suburban [sə'bɜːbn] Vorort-
display [dɪ'spleɪ] Anzeige
satellite ['sætəlaɪt] Satellit
to pinpoint ['pɪnpɔɪnt] genau anzeigen, genau bestimmen
location [ləʊ'keɪʃn] Lage, Standort
to equip [ɪ'kwɪp] ausrüsten, ausstatten mit
GPS [ˌdʒiːpiː'es] (Global Positioning System) [ˌgləʊbl pə'zɪʃənɪŋ ˌsɪstəm] GPS-Navigationssystem
additional [ə'dɪʃənl] zusätzlich
aid [eɪd] Hilfe, Hilfsmittel
to avoid [ə'vɔɪd] vermeiden
to rely on [rɪ'laɪ‿ɒn] sich verlassen auf
to obtain [ɒb'teɪn] bekommen, erhalten
sensor ['sensə] Sensor, Fühler
to install [ɪn'stɔːl] einbauen
to monitor ['mɒnɪtə] überwachen
interval ['ɪntəvl] zeitlicher Abstand, Intervall
available [ə'veɪləbl] verfügbar, vorhanden
feature ['fiːtʃə] Merkmal, Bestandteil
to regulate ['regjəleɪt] regulieren
traffic flow ['træfɪk ˌfləʊ] Verkehrsfluss
slow moving traffic [ˌsləʊmuːvɪŋ 'træfɪk] zäh fließender Verkehr
to suggest [sə'dʒest] vorschlagen
diversion [daɪ'vɜːʃn] Umleitung
to enter ['entə] eingeben (Daten)
voice recognition ['vɔɪs ˌrekəg'nɪʃn] Spracherkennung
to run out of sth. [rʌn 'aʊt əv ˌsʌmθɪŋ] ausgehen, etwas nicht mehr haben
steering wheel ['stɪərɪŋ ˌwiːl] Lenkrad
petrol station ['petrl ˌsteɪʃn] Tankstelle
to programme ['prəʊgræm] programmieren
command [kə'mɑːnd] Befehl, Führung
accent ['æksənt] Akzent
due to [djuː tuː] aufgrund von
road works ['rəʊdˌwɜːks] Straßenbauarbeiten
to take care [teɪk‿'keə] aufpassen
fog [fɒg] Nebel
to rewrite [ˌriː'raɪt] neu schreiben, umschreiben
glance [glɑːns] Blick

C1

thanks to ['θæŋks tuː] dank
commuter [kə'mjuːtə] Pendler(in)
convenient [kən'viːnɪənt] günstig, praktisch, handlich
to get rid of [get 'rɪd əv] loswerden, abschaffen
attractive [ə'træktɪv] reizvoll, anziehend, attraktiv
to provide [prə'vaɪd] sorgen für, liefern, bereitstellen
flexible ['fleksɪbl] flexibel
fixed-route [ˌfɪkst'ruːt] schienengebunden
freedom ['friːdəm] Freiheit
individual [ˌɪndɪ'vɪdʒʊəl] individuell
crazy ['kreɪzɪ] verrückt
planner ['plænə] Planer(in)
electrically-driven [ɪˌlektrɪklɪ 'drɪvən] elektrisch angetrieben
rail [reɪl] Schiene

ecological [ˌiːkəˈlɒdʒɪkl]	ökologisch, umweltfreundlich
inner [ˈɪnə]	Innen-, inner-
peak time [ˈpiːk ˌtaɪm]	Hauptverkehrszeit
underground [ˈʌndəgraʊnd]	Untergrund, U-Bahn
environmentalist [ɪnˌvaɪrənˈmentəlɪst]	Umweltschützer(in)
to make sense [meɪk ˈsens]	einen Sinn ergeben
medium-sized [ˌmiːdjəm ˈsaɪzd]	mittelgroß
expert [ˈekspɜːt]	Experte, Expertin
to estimate [ˈestɪmeɪt]	schätzen
tramway [ˈtræmweɪ]	Straßenbahngleis
sort [sɔːt]	Art, Sorte
unpopular [ˌʌnˈpɒpjʊlə]	unpopulär
to put into practice [ˌpʌt ɪntʊ ˈpræktɪs]	in die Praxis umsetzen, realisieren
progress [ˈprəʊgres]	Fortschritt
to face sth. [ˈfeɪs‿ˌsʌmθɪŋ]	mit etw. konfrontiert sein
heading [ˈhedɪŋ]	Überschrift
in favour of [ɪn ˈfeɪvər‿əv]	zugunsten von

C2

definite [ˈdefɪnɪt]	eindeutig, bestimmt

D1

delay [dɪˈleɪ]	Verspätung
eastbound section [ˌiːstbaʊnd‿ˈsekʃn]	(Fahrbahn) in östlicher Richtung
detour route [ˈdiːtʊə‿ˌruːt]	Umleitung

D2

coaching days [ˈkəʊtʃɪŋ ˌdeɪz]	Postkutschenzeit
crossroads [ˈkrɒsrəʊdz]	Kreuzung
by rail [baɪ ˈreɪl]	mit der Bahn
to operate [ˈɒpəreɪt]	verkehren, bedienen
on top of [ɒn ˈtɒp əv]	zusätzlich zu
via [ˈvaɪə]	über
base [beɪs]	Basis, Stützpunkt, Ausgangspunkt
town hall [ˌtaʊn ˈhɔːl]	Rathaus
continental [ˌkɒntɪˈnentl]	kontinental

Tapescript

to call at [ˈkɔːl æt]	anhalten
diversion sign [daɪˈvɜːʃn ˌsaɪn]	Umleitungsschild
hold-up [ˈhəʊldʌp]	Stau
platform [ˈplætfɔːm]	Gleis
to proceed [prəˈsiːd]	weitergehen, vorgehen
westbound [ˈwestbaʊnd]	westliche Richtung

Unit 6

Starter

waste [weɪst]	Müll, Abfall
plastic [ˈplæstɪk]	Kunststoff, Plastik
municipal [mjuːˈnɪsɪpl]	kommunal, städtisch
landfill [ˈlændfɪl]	Mülldeponie
composting [ˈkɒmpɒstɪŋ]	Kompostierung
incineration [ɪnˌsɪnəˈreɪʃn]	Verbrennung
chart [tʃɑːt]	Tabelle, Schaubild

A1

acid rain [ˌæsɪd‿ˈreɪn]	saurer Regen
ozone hole [ˈəʊzəʊn ˌhəʊl]	Ozonloch
to contribute [kənˈtrɪbjuːt]	beitragen
to refuse [rɪˈfjuːz]	ablehnen, sich weigern
to influence [ˈɪnflʊəns]	beeinflussen; Einfluss
retailer [ˈriːteɪlə]	Einzelhändler(in)
so-called [ˌsəʊˈkɔːld]	so genannt
environmental [ɪnˌvaɪrənˈmentl]	Umwelt-
to turn to [ˈtɜːn tuː]	sich zuwenden
wasteful [ˈweɪstfʊl]	verschwenderisch
recyclable [ˌriːˈsaɪkləbl]	wiederverwertbar
packaging [ˈpækɪdʒɪŋ]	Verpackung
range [reɪndʒ]	Sortiment, Umfang, Reichweite
to meet the growing demand [ˌmiːt‿ðə ˌgrəʊɪŋ dɪˈmɑːnd]	der wachsenden Nachfrage gerecht werden
manufacturer [ˌmænjəˈfæktʃərə]	Hersteller(in)
advertising campaign [ˈædvətaɪzɪŋ kæmˌpeɪn]	Werbekampagne
contribution [ˌkɒntrɪˈbjuːʃn]	Beitrag
consumption [kənˈsʌmpʃn]	Verbrauch
emission [ɪˈmɪʃn]	Emission, Ausstoß

to insulate ['ɪnsjəleɪt]	isolieren
lighting ['laɪtɪŋ]	Beleuchtung
bulb [bʌlb]	(Glüh)birne
rubbish ['rʌbɪʃ]	Abfall, Müll
ton [tʌn]	Tonne
disposal [dɪ'spəʊsl]	Beseitigung, Entsorgung
conscious ['kɒnʃəs]	bewusst
non-returnable [ˌnɒnrɪ'tɜːnəbl]	Einweg-
at all costs [ət ˌɔːl 'kɒsts]	um jeden Preis
returnable [rɪ'tɜːnəbl]	Mehrweg-
in conclusion [ɪn kən'kluːʒn]	abschließend
to emphasize ['emfəsaɪz]	betonen
once again [wʌns‿ə'gen]	noch einmal
responsibility [rɪˌspɒnsə'bɪlətɪ]	Verantwortung
reaction [rɪ'ækʃn]	Reaktion
reduction [rɪ'dʌkʃn]	Verminderung
insulation [ˌɪnsjə'leɪʃn]	Isolierung
emphasis ['emfəsɪs]	Betonung
to manufacture [ˌmænjə'fæktʃə]	herstellen

A2

chemist ['kemɪst]	Chemiker(in)
fuel [fjʊəl]	Brennstoff, Kraftstoff
elsewhere [ˌels'weə]	woanders
improvement [ɪm'pruːvmənt]	Verbesserung
furthermore ['fɜːðəmɔː]	außerdem, ferner
all over the world [ɔːl ˌəʊvə ðə 'wɜːld]	in der ganzen Welt
CFC [ˌsiːef'siː]	FCKW (Fluorchlorkohlenwasserstoff)
direct speech [daɪˌrekt 'spiːtʃ]	direkte Rede
to take the view [teɪk ðə 'vjuː]	der Ansicht sein
ought to ['ɔːt‿tuː]	sollte(n)
to maintain [meɪn'teɪn]	behaupten
disagreement [ˌdɪsə'griːmənt]	abweichende Meinung, Meinungsverschiedenheit

C

(not) any more [nɒt / ˌenɪ 'mɔː]	nicht länger
to recycle [ˌriː'saɪkl]	wiederverwerten
unlikely [ʌn'laɪklɪ]	unwahrscheinlich
to come true [kʌm 'truː]	wahr werden
to re-use [ˌriː'juːz]	wiederverwenden
raw [rɔː]	Roh-, roh
precise [prɪ'saɪs]	genau
to shift [ʃɪft]	verlagern
Friends of the Earth [ˌfrendz‿əv ðɪ‿'ɜːθ]	Name einer Umweltschutzorganisation
approach [ə'prəʊtʃ]	Ansatz, Annäherung
waste management ['weɪst ˌmænɪdʒmənt]	Abfallentsorgung
scrap [skræp]	Schrott, Fetzen
dealer ['diːlə]	Händler(in)
to dump; dump [dʌmp]	(Müll) abladen; Müllkippe
to dismantle [dɪs'mæntl]	zerlegen, demontieren
to code; code [kəʊd]	kodieren; Kode
to separate ['sepəreɪt]	trennen
left [left]	übrig
supplier [sə'plaɪə]	Lieferant(in)
to substitute; substitute ['sʌbstɪtjuːt]	ersetzen; Ersatz
over and over again [ˌəʊvər‿ən ˌəʊvər‿ə'gen]	immer wieder
card [kɑːd]	Karte
to take to pieces [ˌteɪk tʊ 'piːsɪz]	zerlegen
to melt [melt]	schmelzen
summary ['sʌmərɪ]	Zusammenfassung

D1

ecology [ɪ'kɒlədʒɪ]	Ökologie
zinc [zɪŋk]	Zink
ore [ɔː]	Erz
mining ['maɪnɪŋ]	Bergbau
technologist [tek'nɒlədʒɪst]	Techniker(in)
recording [rɪ'kɔːdɪŋ]	Aufnahme
deposit [dɪ'pɒzɪt]	Lagerstätte, Guthaben
to create [kriː'eɪt]	schaffen, verursachen
nature ['neɪtʃə]	Natur
fear [fɪə]	Furcht

D2

coastal ['kəʊstl]	Küsten ...
surface ['sɜːfɪs]	Oberfläche
attraction [ə'trækʃn]	Anziehungskraft
seaside resort ['siːsaɪd rɪˌzɔːt]	Seebad
major road [ˌmeɪdʒə 'rəʊd]	Hauptstraße
tax [tæks]	Steuer

housing estate ['haʊzɪŋ‿ɪˌsteɪt] — Siedlung
primary school ['praɪmərɪ ˌskuːl] — Grundschule
public library [ˌpʌblɪk 'laɪbrərɪ] — Stadtbücherei
MP (Member of Parliament) [ˌem'piː][ˌmembər‿əv 'pɑːləmənt] — Mitglied des Parlaments
concerned [kən'sɜːnd] — betroffen, interessiert
unemployed [ˌʌnɪm'plɔɪd] — arbeitslos
role [rəʊl] — Rolle
issue ['ɪʃuː] — Thema, Frage
reported speech [rɪˌpɔːtɪd 'spiːtʃ] — indirekte Rede

Tapescript

commercial [kə'mɜːʃl] — gewerblich, kommerziell
hotel owner [həʊˌtel 'əʊnə] — Hoteleigentümer

Unit 7

Starter

environmentally-friendly [ɪnˌvaɪərənˌmentəlɪ'frendlɪ] — umweltfreundlich
limited ['lɪmɪtɪd] — begrenzt
fossil ['fɒsl] — fossil
non-polluting [ˌnɒnpə'luːtɪŋ] — umweltverträglich
radioactive [ˌreɪdɪəʊ'æktɪv] — radioaktiv

A1

wealth [welθ] — Reichtum
impression [ɪm'preʃn] — Eindruck
plenty ['plentɪ] — viel, eine Menge
total ['təʊtl] — gesamt
billion ['bɪljən] — Billion (BE), Milliarde (AE)
mankind [mæn'kaɪnd] — Menschheit
trillion ['trɪljən] — Trillion (AE: Billion)
developing country [dɪˌveləpɪŋ 'kʌntrɪ] — Entwicklungsland, Industrieland
availability [əˌveɪlə'bɪlətɪ] — Vorhandensein, Verfügbarkeit
fossil fuels [ˌfɒsl 'fjʊəlz] — fossile Brennstoffe
to predict [prɪ'dɪkt] — vorhersagen
plentiful ['plentɪfʊl] — reichlich, häufig
standard of living [ˌstændəd‿əv 'lɪvɪŋ] — Lebensstandard

global warming [ˌgləʊbl 'wɔːmɪŋ] — weltweiter Temperaturanstieg
industrialized [ɪn'dʌstrɪəlaɪzd] — industrialisiert
source [sɔːs] — Quelle
solar power [ˌsəʊlə 'paʊə] — Sonnenenergie
to transfer [træns'fɜː] — übertragen
significant [sɪg'nɪfɪkənt] — bedeutend, bedeutsam
to harm [hɑːm] — schädigen
to take action [teɪk 'ækʃn] — etwas unternehmen
what … like? [wɒt … 'laɪk] — wie (ist er/sie/es) …?
antonym ['æntənɪm] — Antonym, Wort mit gegensätzlicher Bedeutung
slight [slaɪt] — leicht, schwach, gering
geographical [ˌdʒɪə'græfɪkl] — geografisch

A2

temperature ['temprətʃə] — Temperatur
BTU British Thermal Unit [ˌbiːtiː'juː] [ˌbrɪtɪʃ ˌθɜːml 'juːnɪt] — BTU = brit. Wärmeeinheit
projection [prə'dʒekʃn] — Prognose, Projektion
constant ['kɒnstənt] — ständig, gleichbleibend
to drop [drɒp] — fallen, fallen lassen
stable ['steɪbl] — stabil, dauerhaft
steady ['stedɪ] — stabil, gleichbleibend
substantial [səb'stænʃl] — beträchtlich, erheblich
sharp [ʃɑːp] — scharf, deutlich, steil
gradual ['grædʒʊəl] — allmählich, sanft (ansteigen/abfallen)

C

in the mid [ɪn ðə 'mɪd] — in der Mitte von, Mitte der …
to generate ['dʒenəreɪt] — erzeugen
to join [dʒɔɪn] — sich anschließen
off-the-grid [ˌɒf ðə 'grɪd] — nicht an das Versorgungsnetz angeschlossen
renewable energy [rɪˌnjuːəbl 'enədʒɪ] — erneuerbare Energie
grid [grɪd] — Versorgungsnetz
power line ['paʊə laɪn] — (Stark-)Stromleitung
as a result [əz‿ə rɪ'zʌlt] — folglich
to supply [sə'plaɪ] — versorgen
yearly ['jɪəlɪ] — jährlich
hydro ['haɪdrəʊ] — Wasser
domestic [də'mestɪk] — häuslich
photovoltaic (pv) [ˌfəʊtəʊvɒl'teɪɪk][ˌpiː'viː] — fotoelektrisch

maintenance ['meɪntənəns]	Wartung, Instandhaltung
maker ['meɪkə]	Hersteller(in)
cell [sel]	Zelle
to dig up [dɪg 'ʌp]	(aus)graben
cable ['keɪbl]	Kabel
traditional [trə'dɪʃənl]	traditionell, herkömmlich
power plant ['paʊə ˌplɑːnt]	Kraftwerk
panel ['pænl]	Schalttafel
research ['riːsɜːtʃ]	Forschung
scientist ['saɪəntɪst]	Wissenschaftler(in)
fuel cell ['fjʊəl ˌsel]	Brennstoffzelle
powered ['paʊəd]	angetrieben, betrieben

D2

fast running [ˌfɑːst 'rʌnɪŋ]	schnell fließend
peat bog ['piːt ˌbɒg]	Torfmoor
resources [rɪ'sɔːsɪz][rɪ'zɔːsɪz]	Boden-, Naturschätze, Mittel
machinery [mə'ʃiːnərɪ]	Maschinen

Tapescript

barrel ['bærl]	Fass
crude oil [ˌkruːd'ɔɪl]	Rohöl
decline [dɪ'klaɪn]	Rückgang
demand [dɪ'mɑːnd]	Nachfrage
exploration block [ˌeksplə'reɪʃn ˌblɒk]	Erkundungsplattformen
previous ['priːvɪəs]	vorhergehend
primary energy consumption [ˌpraɪmərɪ‿'enədʒɪ kən,sʌmpʃn]	Primärenergieverbrauch
to remain [rɪ'meɪn]	bleiben
resurgence [rɪ'sɜːdʒəns]	Wiederaufleben
slump [slʌmp]	Einbruch

Unit 8

Starter

aborigine [ˌæbə'rɪdʒənɪ]	austral. Ureinwohner(in)
chancellor [ˌtʃɑːnsələ]	Kanzler(in)

A1

backpacking ['bækˌpækɪŋ]	Rucksackreisen
embassy ['embəsɪ]	Botschaft
valid ['vælɪd]	gültig
tax-file number ['tæksfaɪl ˌnʌmbə]	Steuernummer
cover ['kʌvə]	Deckung (Versicherung)
personal property [ˌpɜːsnl 'prɒpətɪ]	persönliches Eigentum
parasailing ['pærəˌseɪlɪŋ]	Fallschirmsegeln
jumper ['dʒʌmpə]	Pullover
wrinkles ['rɪŋklz]	Falten
carry-on-luggage [ˌkærɪɒn 'lʌgɪdʒ]	Handgepäck
to label ['leɪbl]	beschriften
unattended [ˌʌnə'tendɪd]	unbeaufsichtigt

A2

to dive [daɪv]	tauchen
to hike [haɪk]	wandern

B1

tourist commission ['tʊərɪst kəˌmɪtiː]	Ausschuss für Tourismus
internet address ['ɪntənet əˌdres]	Internetadresse
layout ['leɪaʊt]	Gestaltung
corrections [kə'rekʃnz]	Korrekturen
to register ['redʒɪstə]	registrieren
cartoon figures [kɑːˌtuːn 'fɪgəz]	Karikaturen
to take out [teɪk 'aʊt]	herausnehmen
letter ['letə]	hier: Buchstaben
to recommend [ˌrekə'mend]	empfehlen
interactivity [ˌɪntəræk'tɪvətɪ]	Interaktivität

B2

hunter-gatherer [ˌhʌntə'gæðərə]	Jäger und Sammler
land bridge ['lændˌbrɪdʒ]	Landbrücken
sea-crossing ['siːˌkrɒsɪŋ]	Meeresüberquerung
to refer to as [rɪ'fɜː tʊ‿əz]	bezeichnen als
penal colony ['piːnəl ˌkɒlənɪ]	Strafkolonie
to found [faʊnd]	gründen
fireworks displays ['faɪəwɜːks dɪˌspleɪz]	Feuerwerke

B3

secondary school ['sekəndrɪ ˌskuːl]	weiterführende Schule, Hauptschule
outback ['aʊtbæk]	Hinterland
arrival [ə'raɪvl]	Ankunft
hostel ['hɒstl]	Herberge
cattle ['kætl]	Vieh
fencing ['fensɪŋ]	Einzäunung
chainsaw ['tʃeɪnsɔː]	Kettensäge
basics ['beɪsɪks]	Grundlagen
to shoe [ʃuː]	beschlagen
awesome ['ɔːsəm]	großartig, eindrucksvoll
unforgettable [ˌʌnfə'getəbl]	unvergesslich

C1

caption ['kæpʃn]	Titel, Überschrift
visual elements [ˌvɪʒʊəl 'eləmənts]	Bildelemente
brochure ['brəʊʃə]	Broschüre

C2

kangaroo [ˌkæŋgər'uː]	Känguru
cartoonist [kɑː'tuːnɪst]	Karikaturist
reluctantly [rɪ'lʌktəntlɪ]	widerwillig
privacy ['praɪvəsɪ]	Privatsphäre

C3

pie chart ['paɪˌtʃɑːt]	Tortendiagramm
purpose ['pɜːpəs]	Zweck
to conclude [kən'kluːd]	folgern
pleasure ['pleʒə]	Vergnügen
accommodation [əˌkɒmə'deɪʃn]	Unterkunft
entertainment [ˌentə'teɪnmənt]	Unterhaltung
expense [ɪk'spens]	Ausgaben

D1

voting system ['vəʊtɪŋ ˌsɪstəm]	Wahlsystem
constituency [kən'stɪtjʊənsɪ]	Wahlkreis
former ['fɔːmə]	frühere(r,s), erstgenannte(r,s)
labour ['leɪbə]	Arbeit
conservative [kən'sɜːvətɪv]	Konservative(r)
democrat ['deməkræt]	Demokrat(in)
distribution [ˌdɪstrɪ'bjuːʃn]	Verteilung

D2

to appoint [ə'pɔɪnt]	ernennen
Governor-General [ˌgʌvnə'dʒenrəl]	Generalgouverneur
virtually ['vɜːtʃʊəlɪ]	nahezu, fast
advice [əd'vaɪs]	Rat
parliamentary leader [ˌpɑːlɪmentrɪ 'liːdə]	Parlamentsführer
proposed [prə'pəʊzd]	vorgeschlagen
bills [bɪlz]	hier: Gesetzentwurf
Lower House [ˌləʊə 'haʊs]	Unterhaus
Upper House [ˌʌpə 'haʊs]	Oberhaus
voter ['vəʊtə]	Wähler(in)

Tapescript

absolute majority [ˌæbsəluːt mə'dʒɒrətɪ]	absolute Mehrheit
to be elected [bɪ‿ɪ'lektɪd]	gewählt werden
to be fortunate [bɪ 'fɔːtʃənət]	Glück haben
compulsory [kəm'pʌlsərɪ]	verpflichtend
to eliminate [ɪ'lɪmɪneɪt]	aussondern
preferences ['prefrənsɪz]	Vorlieben, Präferenzen
primary vote [ˌpraɪmərɪ 'vəʊt]	Erstabstimmung
shadow minister ['ʃædəʊ ˌmɪnɪstə]	Schattenminister
solicitor [sə'lɪsɪtə]	Rechtsanwalt
trade union movement [ˌtreɪdˌjuːnjən 'muːvmənt]	Gewerkschaftsbewegung

Unit 9

Starter

to crawl [krɔːl]	kriechen
to scale [skeɪl]	erklimmen
healing ['hiːlɪŋ]	Heilung
angel ['eɪndʒl]	Engel
devil ['devl]	Teufel
shame [ʃeɪm]	Schande

A

crack addict ['kræk‿ˌædɪkt]	Crack-Abhängige(r)
addict ['ædɪkt]	Abhängige(r)

to be unfamiliar with sth. [bɪ‿ˌʌnfə'mɪljə wɪð ˌsʌmθɪŋ]	etw. nicht kennen, mit etw. nicht vertraut sein
to go missing [ˌgəʊ 'mɪsɪŋ]	verschwinden
sighting ['saɪtɪŋ]	Sichten
(well-) acquainted [ˌwel‿əkweɪntɪd]	(gut) bekannt
runaway ['rʌnəweɪ]	Ausreißer(in)
needle in a haystack [ˌniːdl‿ɪn‿ə 'heɪstæk]	Nadel im Heuhaufen
to keep up with [ˌkiːp‿'ʌp wɪð]	Schritt halten mit
flight [flaɪt]	Flucht
accusation [ˌækjʊ'zeɪʃn]	Anschuldigung
self-esteem [ˌselfɪ'stiːm]	Selbstachtung
in care [ɪn 'keə]	in Pflege
homeless ['həʊmləs]	obdachlos
foster care ['fɒstə ˌkeə]	bei Pflegeeltern
painful ['peɪnfʊl]	schmerzhaft
to permit [pə'mɪt]	erlauben

C

Ecstasy ['ekstəsɪ]	Ecstasy (Name einer Droge)
beat [biːt]	hier: Takt
rave [reɪv]	Rave (größere Tanzveranstaltung zu Technomusik)
to start off [stɑːt‿'ɒf]	anfangen, hier: seinen/ihren Ausgangspunkt haben
Yuppie ['jʌpɪ]	Yuppie (junger karrierebewusster großstädtischer Mensch)
to cause [kɔːz]	verursachen
organs ['ɔːgənz]	Organe
dehydration [ˌdiːhaɪ'dreɪʃn]	Austrocknung, Dehydration
ingredients [ɪn'griːdɪənts]	Bestandteile
pleasurable ['pleʒərəbl]	angenehm
drug counsellor ['drʌg‿ˌkaʊnsələ]	Drogenberater(in)
glamorous ['glæmərəs]	schick, glamourös
hopelessness ['həʊpləsnəs]	Hoffnungslosigkeit
to play hard [ˌpleɪ 'hɑːd]	sich kräftig amüsieren
distinction [dɪ'stɪŋkʃn]	Unterscheidung
layabout ['leɪəˌbaʊt]	Faulenzer(in)
severe [sə'vɪə]	schlimm, ernst, streng

D

loneliness ['ləʊnlɪnəs]	Einsamkeit

Tapescript

to disguise [dɪs'gaɪz]	verschleiern, tarnen
lonely ['ləʊnlɪ]	einsam
long-term [ˌlɒŋ'tɜːm]	langfristig

Unit 10

Starter

immigration [ˌɪmɪ'greɪʃn]	Einwanderung
immigrant ['ɪmɪgrənt]	Einwanderer(in)

A

dilemma [dɪ'lemə]	Klemme, Dilemma
unauthorized [ʌn'ɔːθəraɪzd]	unerlaubt
remedy ['remədɪ]	Heilmittel, Mittel
consecutive [kən'sekjətɪv]	folgend
labor pool ['leɪbə ˌpuːl]	Arbeitsmarkt
decade ['dekeɪd]	Jahrzehnt
macroeconomic [ˌmækrəʊiːkə'nɒmɪk]	gesamtwirtschaftlich, makroökonomisch
despite [dɪ'spaɪt]	trotz
trade barriers ['treɪd‿ˌbærɪəz]	Handelsbeschränkungen
consensus view [kənˌsensəs 'vjuː]	übereinstimmende Ansicht
adequate ['ædɪkwət]	ausreichend
far-reaching [ˌfɑː'riːtʃɪŋ]	weitreichend
tax-system ['tæks‿ˌsɪstəm]	Steuersystem
per-capita income [pəˌkæpɪtər‿'ɪnkʌm]	Pro-Kopf-Einkommen
urgent ['ɜːdʒənt]	dringend
World Trade Organisation [ˌwɜːld‿'treɪd ɔːgənaɪˌzeɪʃn]	Welthandelsorganisation
low-cost supplier [ˌləʊkɒst‿sə'plaɪə]	Billiglieferant
head-on [hed'ɒn]	Kopf-an-Kopf
tomato picker [tə'mɑːtəʊ ˌpɪkə]	Tomatenpflücker

B1

Chinese [ˌtʃaɪ'niːz]	chinesisch

B2

native Americans [ˌneɪtɪv‿ə'merɪkənz]	amerikanische Ureinwohner
American Indian [əˌmerɪkən 'ɪndɪən]	amerikanischer Indianer

C

welfare ['welfeə]	Wohl, Wohlfahrt, Sozialhilfe
corn [kɔːn]	Getreide, Mais, Korn
theory ['θɪərɪ]	Theorie
circle ['sɜːkl]	Kreis, Kreislauf
society [sə'saɪətɪ]	Gesellschaft
to confirm [kən'fɜːm]	bestätigen
poll [pəʊl]	Umfrage, Wahl, Abstimmung
worried ['wʌrɪd]	besorgt
native-born [ˌneɪtɪv'bɔːn]	gebürtig
inflow ['ɪnfləʊ]	Zustrom
short-term ['ʃɔːttɜːm]	kurzfristig
economist [ɪ'kɒnəmɪst]	Wirtschaftswissenschaftler(in)
in the long run [ɪn‿ðə 'lɒŋ‿ˌrʌn]	auf lange Sicht
food stamp ['fuːd‿ˌstæmp]	Essensmarke
unemployment benefit [ˌʌnɪm'plɔɪmənt ˌbenɪfɪt]	Arbeitslosenunterstützung
earnings ['ɜːnɪŋz]	Einkommen, Verdienst, Ertrag

D1

climate ['klaɪmət]	Klima
way of life [ˌweɪ‿əv‿'laɪf]	Lebensweise
marital status [ˌmærɪtl 'steɪtəs]	Familienstand

D2

to publish ['pʌblɪʃ]	veröffentlichen, publizieren
editor ['edɪtə]	Herausgeber(in), Redakteur(in)

Tapescript

Cambodia [ˌkæm'bəʊdɪə]	Kambodscha
food-processing company ['fuːdˌprəʊsesɪŋ ˌkʌmpənɪ]	Nahrungsmittelunternehmen
(to be) homesick ['həʊmsɪk]	Heimweh (haben)
jew [dʒuː]	Jude
Khmer Rouge [kəˌmeə'ruːʒ]	Rote Khmer
painter ['peɪntə]	Maler(in)
refugee [ˌrefjʊ'dʒiː]	Flüchtling
Russian ['rʌʃn]	russisch
savings ['seɪvɪŋz]	Ersparnisse
visa ['viːzə]	Visum

Unit 11

1

behaviour [bɪ'heɪvjə]	Verhalten, Benehmen
to get along [get ə'lɒŋ]	gut miteinander auskommen
to pick up [pɪk 'ʌp]	jmdn. oder etwas abholen
meat loaf ['miːtləʊf]	Hackbraten
mashed potatoes [ˌmæʃt‿pə'teɪtəʊz]	Kartoffelpüree
pepper ['pepə]	Pfeffer
can [kæn]	Dose
bathroom ['bɑːθruːm]	Badezimmer, Toilette
impolite [ˌɪmpə'laɪt]	unhöflich
to blow one's nose [ˌbləʊ wʌnz 'nəʊz]	sich die Nase putzen
hall [hɔːl]	Flur
unpractical [ʌn'præktɪkl]	unpraktisch
cold [kəʊld]	Erkältung
pants [pænts]	(AE) lange Hose, (BE) Unterhose
cashier [kæʃ'ɪə]	Kassierer(in)
baptist ['bæptɪst]	Baptist
service ['sɜːvɪs]	Dienst, Dienstleistung, Gottesdienst
religious [rɪ'lɪdʒəs]	religiös
topic ['tɒpɪk]	Thema
contrary to ['kɒntrərɪ tuː]	im Gegensatz zu
to attend [ə'tend]	teilnehmen
worship ['wɜːʃɪp]	Anbetung, Verehrung, Gottesdienst
apparent [ə'pærənt]	offensichtlich
tiny ['taɪnɪ]	winzig
straps [stræps]	Träger
to expose [ɪk'spəʊz]	entblößen
stomach ['stʌmɪk]	Bauch
bandanas [ˌbæn'dænəz]	Halstuch, Kopftuch
PDA (public demonstration of affection) [ˌpiːdiː'eɪ][ˌpʌblɪk ˌdemən'streɪʃn əv ə'fekʃn]	öffentliche Zurschaustellung von Zuneigung
affection [ə'fekʃn]	Zuneigung, Liebe
disciplined ['dɪsəplɪnd]	diszipliniert
grade [greɪd]	Abschluss, Note
senior ['siːnɪə]	Schüler(in) der Abschlussklasse

term [tɜːm]	Semester, Schulhalbjahr
schedule ['ʃedjuːl]['skedjuːl]	Liste, Fahrplan, Stundenplan
astounded [ə'staʊndɪd]	verblüfft, überrascht
Pledge of Allegiance [ˌpledʒ‿əv ə'liːdʒəns]	Treueschwur (auf die Fahne)
it's my turn ['maɪ ˌtɜːn]	ich bin an der Reihe

2

freeway ['friːweɪ]	Autobahn (AE)
soccer ['sɒkə]	Fußball (AE)
baggage ['bægɪdʒ]	Gepäck (AE)
closet ['klɒzɪt]	Schrank (AE)
gas [gæs]	Benzin (AE), Gas
trunk [trʌŋk]	Kofferraum (AE)
cookies ['kʊkɪz]	Kekse, Plätzchen
rest room ['restruːm]	Toilette (AE) (auch restroom)
sidewalk ['saɪdwɔːk]	Bürgersteig (AE)
fall [fɔːl]	Herbst (AE)
vacation [və'keɪʃn][veɪ'keɪʃn]	Ferien, Urlaub (AE)
French fries [ˌfrentʃ 'fraɪz]	Pommes frites (AE)
wardrobe ['wɔːdrəʊb]	Kleiderschrank
Labor Day ['leɪbəˌdeɪ]	Tag der Arbeit
Thanksgiving [ˌθæŋks'gɪvɪŋ]	amerikanisches Erntedankfest
Independence Day [ˌɪndɪ'pendəns ˌdeɪ]	Unabhängigkeitstag
similarity [ˌsɪmɪ'lærətɪ]	Ähnlichkeit

Unit 12

1

cross-cultural [ˌkrɒs'kʌltʃərəl]	interkulturell
accounts department [ə'kaʊnts dɪˌpɑːtmənt]	Buchhaltung
to freshen up [ˌfreʃn 'ʌp]	sich frisch machen

2

side salad ['saɪd ˌsæləd]	Salat zum Hauptgericht
cod [kɒd]	Kabeljau
peas [piːz]	Erbsen
Masala chicken [mæˌsɑːlə 'tʃɪkɪn]	Hühnchen nach indischer Art
apple pie [ˌæpl 'paɪ]	Art Apfelstrudel
custard ['kʌstəd]	Vanille(soße)

4

to hug [hʌg]	in Arm nehmen
eye contact ['aɪ ˌkɒntækt]	Blickkontakt

5

negotiator [nɪ'gəʊʃɪeɪtə]	Verhandlungspartner(in)
potential client [pəˌtenʃl 'klaɪənt]	möglicher Kunde
proposal [prə'pəʊzl]	Angebot
etiquette ['etɪket]	Etikette
approach (to business) [əˌprəʊtʃ tʊ 'bɪznɪs]	Herangehensweise bei einem Geschäft
to close a deal [ˌkləʊz ə 'diːl]	Geschäft abschließen
proceedings [prə'siːdɪŋz]	Abläufe
to convey [kən'veɪ]	übermitteln
confidence ['kɒnfɪdəns]	Zuversicht, Vertrauen
sincerity [sɪn'serətɪ]	Ehrlichkeit
trustworthiness ['trʌstˌwɜːðɪnəs]	Treue, Vertrauen
prolonged [prə'lɒŋd]	anhaltend
rude [ruːd]	unhöflich
tactile ['tæktaɪl]	Körperkontakt mögen
gap [gæp]	Lücke
to converse [kən'vɜːs]	sich unterhalten
"clock conscious" ['klɒk‿ˌkɒnʃəs]	hier: mit Sinn für Pünktlichkeit
crucial ['kruːʃl]	entscheidend, äußerst wichtig
insult ['ɪnsʌlt]	Beleidigung
urgency ['ɜːdʒənsɪ]	Dringlichkeit
gender ['dʒendə]	Geschlecht
to perceive [pə'siːv]	empfinden, wahrnehmen, erkennen
to address people [ə'dres ˌpiːpl]	jemanden anreden
gift [gɪft]	Geschenk
integral part [ˌɪntɪgrəl 'pɑːt]	integraler Bestandteil
business protocol [ˌbɪznɪs 'prəʊtəkɒl]	Business-Etikette
connotation [ˌkɒnə'teɪʃn]	Bedeutung
lavish (gifts) ['lævɪʃ]	üppig, großzügig
to reciprocate (gifts) [rɪ'sɪprəkeɪt]	abwechselnd wirken, austauschen
to wrap [ræp]	einwickeln
to impact [ɪm'pækt]	beeinflussen
foundation [faʊn'deɪʃn]	Stiftung

Unit 13

1

to seek [siːk]	suchen
individuals [ˌɪndɪ'vɪdjʊəlz]	Personen
Continent ['kɒntɪnənt]	Kontinent
Manchester-based [ˌmæntʃestə'beɪst]	in Manchester ansässig
A level ['eɪˌlevl]	etwa: Abitur
equivalent [ɪ'kwɪvələnt]	entsprechend
skills [skɪlz]	Fähigkeiten
native speaker competence [ˌneɪtɪvˌspiːkə 'kɒmpətəns]	muttersprachliche Kompetenz
initial [ɪ'nɪʃl]	anfänglich
comprehensive [ˌkɒmprɪ'hensɪv]	umfassend
rates of pay [ˌreɪts əv 'peɪ]	Bezahlung
challenging ['tʃæləndʒɪŋ]	herausfordernd
permanent position [ˌpɜːmənənt pə'zɪʃn]	Festanstellung
recruitment officer [rɪ'kruːtmənt ˌɒfɪsə]	Verantwortlicher für die Personalbeschaffung
travel reservations ['trævl rezəˌveɪʃnz]	Reisebuchung
reservations agent [ˌrezə'veɪʃnz ˌeɪdʒnt]	Mitarbeiter(in) in der Reisereservierung
chartered flight [ˌtʃɑːtəd 'flaɪt]	Charterflug
scheduled flight [ˌʃedjuːld 'flaɪt]	Linienflug
car hire ['kɑː ˌhaɪə]	Autovermietung
insurance [ɪn'ʃɔːrəns]	Versicherung
add-ons ['ædɒnz]	Zusatzdienste
customer service [ˌkʌstəmə 'sɜːvɪs]	Kundenservice
pro-actively [ˌprəʊ'æktɪvlɪ]	mit Eigeninitiative
fluency ['fluːənsɪ]	Sprachbeherrschung
command [kə'mɑːnd]	hier: Sprachbeherrschung
covering letter (=letter of application) ['kʌvərɪŋ ˌletə]	Anschreiben
apprenticeship [ə'prentɪʃɪp]	Ausbildung
traineeship (BE) [ˌtreɪ'niːʃɪp]	Praktikum
to state [steɪt]	nennen
briefly ['briːflɪ]	kurz

2

selling tool ['selɪŋ ˌtuːl]	Verkaufswerkzeug
to outline ['aʊtlaɪn]	zusammenfassen
prospective [prə'spektɪv]	zukünftig
finally ['faɪnəlɪ]	schließlich
to address [ə'dres]	adressieren an
hiring decision ['haɪrɪŋ dɪˌsɪʒn]	Einstellungsentscheidung
to individualise [ˌɪndɪ'vɪdjʊəlaɪz]	individualisieren
frowned upon ['fraʊnd əˌpɒn]	verpönt
impersonal [ɪm'pɜːsənəl]	unpersönlich
salutation [ˌsæljʊ'teɪʃn]	Anrede
Dear Sir or Madam ['dɪə ˌsɜːr‿ɔː 'mædəm]	Sehr geehrte Damen und Herren
divorced [dɪ'vɔːst]	geschieden
form of address [ˌfɔːm əv ə'dres]	Anrede
Management Assistant in Informatics [ˌmænɪdʒmənt ə'sɪstənt ɪn ˌɪnfə'mætɪks]	Informatikkaufmann/ -kauffrau
Management Assistant in Event Organisation [ˌmænɪdʒmənt ə'sɪstənt ɪn ɪ'vent ɔːgənaɪˌzeɪʃn]	Veranstaltungskaufmann/ -kauffrau
Management Assistant in Hotel and Hospitality [ˌmænɪdʒmənt ə'sɪstənt ɪn həʊˌtel ənd hɒspɪ'tælɪtɪ]	Hotelkaufmann/ -kauffrau
Management Assistant in Warehousing and Logistics [ˌmænɪdʒmənt ə'sɪstənt ɪn ˌweəhaʊzɪŋ ənd lə'dʒɪstɪks]	Fachkraft für Lagerwirtschaft
Automobile Sales Management Assistant [ˌɔːtəməʊbiːl 'seɪlz ˌmænɪdʒmənt ə'sɪstənt]	Automobilkaufmann/ -kauffrau
Management Assistant in Sports and Fitness [ˌmænɪdʒmənt ə'sɪstənt ɪn ˌspɔːts ən 'fɪtnəs]	Sport- und Fitnesskaufmann/ -kauffrau
Travel Management Assistant ['trævl ˌmænɪdʒmənt ə'sɪstənt]	Reiseverkehrskaufmann/ -kauffrau

Management Assistant in Retail Business [ˌmænɪdʒmənt əˈsɪstənt ɪn ˌriːteɪl ˈbɪznɪs]	Kaufmann/-frau im Einzelhandel
Bank Business Management Assistant [ˈbæŋk ˌbɪznɪs ˌmænɪdʒmənt əˈsɪstənt]	Bankkaufmann/-kauffrau
Office Management Assistant [ˈɒfɪs ˌmænɪdʒmənt əˈsɪstənt]	Bürokaufmann/-kauffrau

3

aware [əˈweə]	bewusst
profession [prəˈfeʃn]	Beruf
examination [ɪgˌzæmɪˈneɪʃn]	Prüfung
to enable [ɪˈneɪbl]	ermöglichen
to enrol [ɪnˈrəʊl]	sich immatrikulieren
polytechnic university (BE) [pɒlɪˌteknɪk‿juːnɪˈvɜːsətɪ]	Fachhochschule
overall grade [ˌəʊvərɔːl ˈgreɪd]	Durchschnittsnote
intern [ˈɪntɜːn]	Praktikant(in)
temporary employment [ˌtemprərɪ‿ɪmˈplɔɪmənt]	Anstellung mit Zeitvertrag
to sign [saɪn]	unterschreiben
to paraphrase [ˈpærəfreɪz]	umschreiben
diplomas [dɪˈpləʊməz]	Diplome, Prüfungsurkunden
valuable [ˈvæljʊbl]	wertvoll
to desire [dɪˈzaɪə]	wünschen
race [reɪs]	Rasse

4

to commit [kəˈmɪt]	hier: investieren
properly [ˈprɒpəlɪ]	richtig
recent [ˈrɪːsənt]	neueste(r,s)
achievement [əˈtʃiːvmənt]	Erfolg
self-confidence [ˌselfˈkɒnfɪdəns]	Selbstvertrauen
to evaluate [ɪˈvæljʊeɪt]	beurteilen
weakness [ˈwiːknəs]	Schwäche
to elaborate [ɪˈlæbəreɪt]	näher ausführen
potential [pəˈtenʃl]	möglich
impressive [ɪmˈpresɪv]	beeindruckend
communication skills [kəˌmjuːnɪˈkeɪʃn ˌskɪlz]	Kommunikationsfähigkeit
beforehand [bɪˈfɔːhænd]	zuvor
soft skills [ˈsɒft ˌskɪlz]	Soziale Kompetenz
professional [prəˈfeʃənl]	professionell
neat [niːt]	gepflegt
appearance [əˈpɪərəns]	Erscheinungsbild
polished [ˈpɒlɪʃt]	poliert
hairdo [ˈheəduː]	Frisur
simply tailored [ˌsɪmplɪ ˈteɪləd]	einfach geschnitten
suit [sjuːt]	Anzug / Kostüm
pressed [prest]	gebügelt
flashy [ˈflæʃɪ]	auffallend
jewellery [ˈdʒuːəlrɪ]	Schmuck
vocational school (part-time) [vəʊˈkeɪʃənl ˌskuːl] [ˌpɑːtˈtaɪm]	Berufsschule
higher secondary vocational school [ˈhaɪə ˌsekəndrɪ vəʊˈkeɪʃənl ˌskuːl]	Fachoberschule, Berufskolleg
secondary modern school [ˌsekəndrɪ ˈmɒdən ˌskuːl]	Hauptschule
higher secondary modern school [ˈhaɪə ˌsekəndrɪ ˈmɒdən ˌskuːl]	Realschule
grammar school [ˈgræmə ˌskuːl]	Gymnasium
vocational school (full-time) [vəʊˈkeɪʃənl ˌskuːl] [ˌfʊlˈtaɪm]	Berufsfachschule

Unit 14

1

recently [ˈriːsəntlɪ]	kürzlich
Far East [ˌfɑːr‿ˈiːst]	Ferner Osten
dynamic [daɪˈnæmɪk]	dynamisch
industrial region [ɪnˈdʌstrɪəl ˌriːdʒən]	Industrieregion
to build up [bɪld ˈʌp]	aufbauen
solid [ˈsɒlɪd]	fest, solide
network [ˈnetwɜːk]	Netzwerk
partnership [ˈpɑːtnəʃɪp]	Partnerschaft, Personengesellschaft
to benefit [ˈbenɪfɪt]	profitieren
unrivalled [ʌnˈraɪvld]	unübertroffen
household goods [ˌhaʊshəʊld‿ˈgʊdz]	Haushaltswaren
to be involved in [bɪ‿ɪnˈvɒlvd ɪn]	an etwas beteiligt sein

headquarters [ˌhed'kwɔːtəz]	Hauptquartier
liability [ˌlaɪə'bɪlətɪ]	Haftung
ownership ['əʊnəʃɪp]	Eigentum

2

secretarial staff [sekrəˌteərɪəl 'stɑːf]	Büroangestellte
to arrange [ə'reɪndʒ]	ausmachen, arrangieren, zuwege bringen
appointment [ə'pɔɪntmənt]	Termin
voicemail ['vɔɪsmeɪl]	Anrufbeantworter
to check [tʃek]	überprüfen
to tend [tend]	tendieren
sales department ['seɪlz dɪˌpɑːtmənt]	Verkaufsabteilung
complaint [kəm'pleɪnt]	Beschwerde
(to be) off sick [ˌɒf 'sɪk]	krankgeschrieben (sein)
client ['klaɪənt]	Kunde

3

to reject [rɪ'dʒekt]	zurückweisen
to suit [sjuːt]	passen
to be engaged [bɪ‿ɪn'geɪdʒd]	verplant sein
to cancel ['kænsl]	absagen
to postpone [pəʊs'pəʊn]	verschieben
diary ['daɪərɪ]	Terminplan

4

non-stop [ˌnɒn'stɒp]	ohne Zwischenlandung
one way ticket [ˌwʌnweɪ 'tɪkɪt]	Fahrschein für einfache Fahrt
return ticket [ˌrɪtɜːn 'tɪkɪt]	Fahrschein für Hin- und Rückfahrt
non-smoking compartment [ˌnɒn'sməʊkɪŋ‿kəmˌpɑːtmənt]	Nichtraucherabteil
at the latest [ət‿ðə 'leɪtɪst]	spätestens
aisle seat ['aɪl ˌsiːt]	Gangplatz
departure [dɪ'pɑːtʃə]	Abflug, Abreise
desk [desk]	hier: Schalter
flight schedule ['flaɪt‿ˌʃedjuːl]	Flugplan

5

business associate ['bɪznɪs əˌsəʊsɪət]	Geschäftspartner(in)
single room [ˌsɪŋgl 'ruːm]	Einzelzimmer
double room [ˌdʌbl 'ruːm]	Doppelzimmer
half board [ˌhɑːf'bɔːd]	Halbpension
full board [ˌfʊl'bɔːd]	Vollpension
facilities [fə'sɪlətɪz]	Einrichtungen
valet parking [ˌvæleɪ 'pɑːkɪŋ]	Auto des Gastes wird eingeparkt
disabled [dɪ'seɪbld]	behindert
receptionist [rɪ'sepʃənɪst]	Rezeptionist(in)
to draw up [drɔː‿'ʌp]	aufstellen
superior room [suːˌpɪərɪə‿'ruːm]	Luxuszimmer
twin beds [ˌtwɪn 'bedz]	zwei Einzelbetten
double bed [ˌdʌbl 'bed]	Doppelbett
to accommodate [ə'kɒmədeɪt]	unterbringen
overhead projector [ˌəʊvəhed prə'dʒektə]	Tageslicht-Projektor
video projector [ˌvɪdɪəʊ prə'dʒektə]	Beamer
to regret [rɪ'gret]	bedauern

6

netiquette ['netɪket]	Umgangsformen im Internet
to behave [bɪ'heɪv]	sich benehmen
subject title ['sʌbdʒɪkt‿ˌtaɪtl]	Betreff
chain letter ['tʃeɪn ˌletə]	Kettenbrief
mixed case [ˌmɪkst 'keɪs]	Groß- und Kleinschreibung
unsolicited [ˌʌnsə'lɪsɪtɪd]	unerbeten
to attach [ə'tætʃ]	anhängen
to capitalize ['kæpɪtəlaɪz]	groß schreiben
sarcasm ['sɑːkæzm]	Sarkasmus
humour ['hjuːmə]	Humor
forbidden [fə'bɪdn]	verboten

7

purchasing ['pɜːtʃəsɪŋ] representative [ˌreprɪ'zentətɪv]	Einkaufs-Vertreter(in)
mutual ['mjuːtʃʊəl]	gegenseitig, gemeinsam
particulars [pə'tɪkjələz]	Einzelheiten
confident [kɒnfɪdənt]	zuversichtlich
to display [dɪ'spleɪ]	zeigen, anzeigen
item ['aɪtəm]	Punkt, Posten, Gegenstand, Artikel

agenda [ə'dʒendə]	Tagesordnung, Programm
doubtless ['daʊtləs]	zweifellos
grateful ['greɪtfʊl]	dankbar
delighted [dɪ'laɪtɪd]	erfreut
factual ['fæktʃʊəl]	sachlich
to the point [tə ðə 'pɔɪnt]	treffend, zur Sache gehörig
correspondence [ˌkɒrɪ'spɒndəns]	Korrespondenz, Briefverkehr
request [rɪ'kwest]	Bitte, Anliegen, Anfrage
appropriate [ə'prəʊprɪət]	angemessen, angebracht

8

conference room ['kɒnfrəns ˌruːm]	Konferenzzimmer
bumpy ['bʌmpɪ]	holprig, unruhig
refreshments [rɪ'freʃmənts]	Erfrischungen
bright [braɪt]	hell
actually ['æktʃʊəlɪ]	tatsächlich
staff [stɑːf]	Belegschaft, Mitarbeiter

9

reception desk [rɪ'sepʃn ˌdesk]	Rezeption, Empfangsschalter
to direct [daɪ'rekt]	leiten
floorplan ['flɔːˌplæn]	Grundriss
stairs [steəz]	Treppen
first floor [ˌfɜːst'flɔː]	Erdgeschoss (AE), 1. Obergeschoss (BE)
ground floor (BE) [ˌgraʊnd'flɔː]	Erdgeschoss
Marketing and Export Department [ˌmɑːkɪtɪŋ ən ' ekspɔːt‿dɪˌpɑːtmənt]	Marketing und Exportabteilung
storage ['stɔːrɪdʒ]	Lager
Marketing Manager ['mɑːkɪtɪŋ ˌmænɪdʒə]	Abteilungsleiter Marketing
Gents [dʒents]	Herren (WC)
open plan office [ˌəʊpnplæn 'ɒfɪs]	Großraumbüro
Managing Director [ˌmænɪdʒɪŋ daɪ'rektə]	Geschäftsführer(in)
Research Manager ['riːsɜːtʃ ˌmænɪdʒə]	Abteilungsleiter(in) Forschung
Export Manager ['ekspɔːt ˌmænɪdʒə]	Abteilungsleiter(in) Export
Data Processing ['deɪtə ˌprəʊsesɪŋ]	Datenverarbeitung
Finance and Personnel Department [ˌfaɪnæns ən pɜːsən'el dɪˌpɑːtmənt]	Finanz- und Personalabteilung
Company Training Manager [ˌkʌmpənɪ 'treɪnɪŋ ˌmænɪdʒə]	Ausbildungsleiter(in)
Chief Accountants [ˌtʃiːf ə'kaʊntənt]	Hauptbuchhalter
canteen [kæn'tiːn]	Kantine
stationary ['steɪʃənrɪ]	Bürobedarf
lounge area ['laʊndʒ ˌeərɪə]	hier: Halle

10

menu ['menjuː]	Speisekarte
main course [ˌmeɪn 'kɔːs]	Hauptgericht
starter ['stɑːtə]	Vorspeise
dessert [dɪ'zɜːt]	Nachtisch
separate ['seprət]	getrennt
change [tʃeɪndʒ]	hier: Wechselgeld
to fetch [fetʃ]	holen, abholen
salmon ['sæmən]	Lachs
soup of the day [ˌsuːp əv ðə 'deɪ]	Tagessuppe
Here you go. [ˌhɪə jə 'gəʊ]	Bitte schön.
tip [tɪp]	Trinkgeld

Tapescript

country code ['kʌntrɪ ˌkəʊd]	Landesvorwahl
extension [ɪk'stenʃn]	Durchwahl
kitchen equipment ['kɪtʃɪn ɪˌkwɪpmənt]	Küchengeräte

Unit 15

A

stereotype ['sterɪəʊtaɪp]	Stereotyp, Klischee
middle-class [ˌmɪdl'klɑːs]	Mittelklasse
to insist (on) [ɪn'sɪst‿ɒn]	beharren, bestehen (auf)
sex [seks]	Geschlechtsverkehr, Geschlecht
birth control ['bɜːθkənˌtrəʊl]	Geburtenkontrolle
to live on welfare [ˌlɪv ɒn 'welfeə]	von Sozialhilfe leben
survival [sə'vaɪvl]	Überleben

whatever [wɒt'evə]	was auch immer
to come to hand [kʌm tʊ 'hænd]	zur Verfügung stehen
junk [dʒʌŋk]	Gerümpel, Ramsch
open air [ˌəʊpn'eə]	Freiluft-, draußen
illusion [ɪ'lu:ʒn]	Illusion, Einbildung
gambler ['gæmblə]	Spieler(in)
alcoholic [ˌælkə'hɒlɪk]	Alkoholiker(in)
to sign up for [saɪn‿'ʌp fɔ:]	sich einschreiben, anmelden
loaf of bread [ˌləʊf‿əv 'bred]	Laib Brot
kindness ['kaɪndnəs]	Güte, Freundlichkeit
charity ['tʃærətɪ]	karitative Organisation, Wohlfahrt
expectation [ˌekspek'teɪʃn]	Erwartung
deprivation [ˌdeprɪ'veɪʃn]	Entbehrung, Entzug
genius ['dʒi:njəs]	Genie
to make do [meɪk 'du:]	auskommen
second hand [ˌsekənd'hænd]	aus zweiter Hand, gebraucht
uneducated [ˌʌn'edjʊkeɪtɪd]	ungebildet
church-run [ˌtʃɜ:tʃ'rʌn]	kirchlich
to declare [dɪ'kleə]	verkünden
kid [kɪd]	Kind, Jugendlicher
to mind one's own business [ˌmaɪnd wʌnz‿əʊn 'bɪznɪs]	sich um seine eigenen Angelegenheiten kümmern
to abuse; abuse [ə'bju:z][ə'bju:s]	missbrauchen; Missbrauch
to misuse; misuse [ˌmɪs'ju:z][ˌmɪs'ju:s]	missbrauchen; Missbrauch
vast [vɑ:st]	riesig, weit
recipient [rɪs'ɪpɪənt]	Empfänger(in)
dependent [dɪ'pendənt]	abhängig, hier: minderjährig

B1

fault [fɔ:lt] [fɒlt]	Schuld
discotheque ['dɪskətek]	Disko
pregnant ['pregnənt]	schwanger

B2

nativity [nə'tɪvətɪ]	Geburt, hier: Herkunft

C1

material [mə'tɪərɪəl]	materiell
damp [dæmp]	feucht
inadequate [ɪn'ædɪkwət]	nicht ausreichend, unangemessen
cultural ['kʌltʃərəl]	kulturell
encouragement [ɪn'kʌrɪdʒmənt]	Ermutigung
gang [gæŋ]	Bande
clothing ['kləʊðɪŋ]	Kleidung
peer group ['pɪəgru:p]	Gruppe von Gleichgesinnten
poorly paid [ˌpɔ:lɪ 'peɪd]	schlecht bezahlt
unstable [ʌn'steɪbl]	unstabil

C2

voluntary organization [ˌvɒləntrɪ‿ˌɔ:gənaɪ'zeɪʃn]	freiwillige Organisation

Tapescript

abortion [ə'bɔ:ʃn]	Abtreibung
to beg [beg]	betteln
binge drinking ['bɪndʒˌdrɪŋkɪŋ]	Besäufnis, Saufgelage
to calm down [kɑ:m 'daʊn]	beruhigen
to miss out on [ˌmɪs 'aʊt ɒn]	etwas verpassen
to throw out [θrəʊ‿'aʊt]	entlassen, feuern
tranquilliser ['træŋkwɪlaɪzə]	Beruhigungsmittel

Unit 16

A

teenage ['ti:neɪdʒ]	im Teenageralter
confused [kən'fju:zd]	verwirrt
to play truant [pleɪ 'tru:ənt]	(die Schule) schwänzen
shoplifting ['ʃɒpˌlɪftɪŋ]	Ladendiebstahl
judge [dʒʌdʒ]	Richter(in)
moral ['mɒrəl]	Moral; moralisch
crime [kraɪm]	Verbrechen
crime level ['kraɪm ˌlevl]	Verbrechensrate
indicator ['ɪndɪkeɪtə]	Anzeiger, Indikator
drug [drʌg]	Droge
immoral [ɪ'mɒrəl]	unmoralisch
violent ['vaɪələnt]	gewalttätig
deprived [dɪ'praɪvd]	unterprivilegiert
acceptable [ək'septəbl]	akzeptabel, annehmbar
income support [ˌɪnkʌm sə'pɔ:t]	Sozialunterstützung
to abolish [ə'bɒlɪʃ]	abschaffen

youth training place [ˌjuːθ 'treɪnɪŋ ˌpleɪs]	Ausbildungsplatz für Jugendliche
unfulfilled [ˌʌnfʊl'fɪld]	unerfüllt
suicide ['suːɪsaɪd]	Selbstmord
alcohol ['ælkəhɒl]	Alkohol
likely ['laɪklɪ]	wahrscheinlich
victim ['vɪktɪm]	Opfer
variety [və'raɪətɪ]	Auswahl
hooligan ['huːlɪgən]	Hooligan, Randalierer
respectable [rɪ'spektəbl]	angesehen, anständig
to rebel; rebel [rɪ'bel]['rebl]	rebellieren; Rebell
rose-tinted glasses [ˌrəʊztɪntɪd 'glɑːsɪz]	rosarote Brille
amnesia [æm'niːzɪə]	Amnesie, Gedächtnisverlust
forgetfulness [fə'getfʊlnəs]	Vergesslichkeit
to panic; panic ['pænɪk]	in Panik geraten, in Panik versetzen; Panik
to shock; shock [ʃɒk]	schockieren; Schock

B1

offender [ə'fendə]	Straftäter(in)

C1

case study ['keɪsˌstʌdɪ]	Fallstudie
joyride ['dʒɔɪraɪd]	(wilde) Vergnügungsfahrt
scrap yard ['skræpjɑːd]	Schrottplatz
movie ['muːvɪ]	Film

C2

insurance fraud [ɪn'ʃʊərəns ˌfrɔːd]	Versicherungsschwindel

Tapescript

to smash [smæʃ]	zerschlagen
straight people [ˌstreɪt 'piːpl]	normale Leute

Unit 17

A

to pass away [pɑːs‿ə'weɪ]	entschlafen, hinscheiden; Umschreibung für: sterben
Holy Bible [ˌhəʊlɪ 'baɪbl]	(Heilige) Bibel
evolution [ˌiːvə'luːʃn]	Entwicklung, Evolution
for better or for worse [fə ˌbetər‿ɔː fə 'wɜːs]	zum Besseren oder Schlechteren hin
smokestack ['sməʊkstæk]	Schornstein
harmful ['hɑːmfʊl]	schädlich
atmosphere ['ætməsfɪə]	Atmosphäre
stream [striːm]	Bach, Wasserlauf
irreplaceable [ˌɪrɪ'pleɪsəbl]	unersetzlich
forest ['fɒrɪst]	Wald
to cut [kʌt]	reduzieren
to poison; poison ['pɔɪzən]	vergiften; Gift
pesticide ['pestɪsaɪd]	Pestizid
chemical ['kemɪkl]	chemisch
ozone layer ['əʊzəʊn‿ˌleɪə]	Ozonschicht
chlorofluorocarbon (CFC) [ˌklɔːrəʊˌflʊərəʊ'kɑːbən] [ˌsiː ef'siː]	Fluorchlorkohlenwasserstoff (FCKW)
foam [fəʊm]	Schaum
air-conditioner ['eəkənˌdɪʃənə]	Klimaanlage
to threaten ['θretn]	(be-)drohen
threat [θret]	(Be-)Drohung
tropical ['trɒpɪkl]	tropisch
species ['spiːsiːz]	(Tier-, Pflanzen-) Art, Spezies
extinction [ɪk'stɪŋkʃn]	Aussterben
dinosaur ['daɪnəsɔː]	Dinosaurier
consequence ['kɒnsɪkwəns]	Folge, Konsequenz
to pay attention [peɪ‿ə'tenʃn]	aufmerksam sein
centigrade ['sentɪgreɪd]	Celsius
greenhouse effect ['griːnhaʊs‿ɪˌfekt]	Treibhauseffekt, globale Erwärmung
lack [læk]	Mangel
flame [fleɪm]	Flamme
garbage ['gɑːbɪdʒ]	Abfall, Müll
sewage ['suːɪdʒ]	Abwasser
medical ['medɪkl]	ärztlich, medizinisch
bather ['beɪðə]	Badende(r)
to confront [kən'frʌnt]	gegenüberstellen, konfrontieren
Mediterranean (Sea) [ˌmedɪtə'reɪnɪən]	Mittelmeer
English Channel [ˌɪŋglɪʃ'tʃænl]	Ärmelkanal

seal [siːl] Seehund
immune system [ɪ'mjuːn ˌsɪstəm] Immunsystem
to weaken ['wiːkən] schwächen
CO_2 [ˌsiːəʊ'tuː] Kohlendioxyd
to flood [flʌd] überfluten
to ruin ['ruːɪn] ruinieren
infertile [ɪn'fɜːtaɪl] unfruchtbar
uninhabitable [ˌʌnɪn'hæbɪtəbl] unbewohnbar
migration [maɪ'greɪʃn] (Ab-)Wanderung
to halt; halt [hɒlt] anhalten; Halt, Pause, Unterbrechung
massive ['mæsɪv] massiv, wuchtig, enorm
mobilization [ˌməʊbɪlaɪ'zeɪʃn] Mobilisierung
to sacrifice; sacrifice ['sækrɪfaɪs] opfern; Opfer
wartime ['wɔːtaɪm] Kriegszeit
humanity [hjuː'mænətɪ] Menschheit, Menschlichkeit
exaggeration [ɪgˌzædʒə'reɪʃn] Übertreibung
to ally; ally [ə'laɪ]['ælaɪ] verbünden; Verbündete(r)
attempt [ə'temt] Versuch
unless [ʌn'les] wenn … nicht, es sei denn
ordinary ['ɔːdɪnrɪ] gewöhnlich
treatment ['triːtmənt] Behandlung

B1

document shred ['dɒkjəmənt ˌʃred] Papierwolle
food wrappers ['fuːd ˌræpəz] Verpackung von Nahrungsmitteln
to lend [lend] ausleihen
interest rates ['ɪntrəstˌreɪts] Zinsen
advanced [əd'vɑːnst] fortschrittlich
government regulation [ˌgʌvənmənt regjə'leɪʃn] Vorschriften der Regierung

C

to replace [rɪ'pleɪs] ersetzen
out-of-date [ˌaʊt‿əv 'deɪt] veraltet, überholt, altmodisch
substitution [ˌsʌbstɪ'tjuːʃn] Ersetzen, Austausch
landfill deposit ['lændfɪl dɪˌpɒzɪt] Mülldeponie
soil [sɔɪl] Boden, Erde
avoidance [ə'vɔɪdəns] Vermeidung
physical ['fɪzɪkl] physikalisch, physisch
to recharge [ˌriː'tʃɑːdʒ] wieder aufladen
to treat [triːt] behandeln
cardboard packaging [ˌkɑːdbɔːd 'pækɪdʒɪŋ] Kartonverpackung

Tapescript

delicate ['delɪkət] heikel
mother nature [ˌmʌðə'neɪtʃə] Mutter Natur
consumerism [kən'sjuːmərɪzm] Konsumdenken
momentum [mə'mentəm] Schwungkraft
awareness [ə'weənəs] Erkenntnis
to dispose of [dɪ'spəʊz‿əv] entsorgen
waxed paper [ˌwækst‿'peɪpə] Wachspapier
carbon paper [ˌkɑːbən 'peɪpə] Kohlepapier
plastic laminated material [ˌplæstɪk ˌlæmɪneɪtɪd mə'tɪərɪəl] kunststoffbeschichtetes Material
to wane [weɪn] schwinden, abnehmen
complacent [kəm'pleɪsənt] selbstzufrieden
cognizant ['kɒgnɪzənt] bewusst
to grasp [grɑːsp] begreifen
governmental agencies [gʌvənˌmentl 'eɪdʒənsɪz] Regierungsstellen
to step up [step 'ʌp] beschleunigen
tax incentives ['tæks‿ɪnˌsentɪvz] Steueranreiz
low-cost loans [ˌləʊkɒst‿'ləʊnz] günstige Kredite
enzymes ['enzaɪmz] Enzyme
ink [ɪŋk] Tinte
to expend [ɪk'spend] aufwenden
to incur [ɪn'kɜː] übernehmen
surplus ['sɜːpləs] überzählig
to lag [læg] hinterherhinken
outpouring ['aʊtˌpɔːrɪŋ] Ausströmen
stumbling block ['stʌmblɪŋ ˌblɒk] Stolperstein
to impede [ɪm'piːd] verhindern, behindern
inability [ˌɪnə'bɪlətɪ] Unvermögen
to exceed [ɪk'siːd] übersteigen
virgin materials [ˌvɜːdʒɪn mə'tɪərɪəlz] ungebrauchte Materialien

Unit 18

A

auction [ˈɔːkʃən]	Auktion
dot.com [dɒtˈkɒm]	Dotcom- (Firma), Online-(Geschäfte)
crisis [ˈkraɪsɪs]	Krise
stock exchange [ˈstɒk‿ɪks,tʃeɪndʒ]	Börse
analyst [ˈænəlɪst]	Analyst(in)
to distinguish [dɪˈstɪŋgwɪʃ]	unterscheiden
rare [reə]	selten
maximum [ˈmæksɪməm]	maximal, Höchst-
interaction [ˌɪntərˈækʃn]	Interaktion
participation [paːˌtɪsɪˈpeɪʃn]	Beteiligung, Teilnahme
tough [tʌf]	hier: schwierig, hart
to hunt [hʌnt]	jagen
bargains [ˈbɑːgɪnz]	günstiges Angebot, Sonderangebot
veteran car [ˈvetərən ˈkaːʳ]	Oldtimer
precious [ˈpreʃəs]	wertvoll, kostbar
to make one's living [ˌmeɪk wʌnz‿ˈlɪvɪŋ]	seinen Lebensunterhalt verdienen
to observe [əbˈzɜːv]	bemerken, feststellen
survivor [səˈvaɪvəʳ]	Überlebende(r)
graduate student [ˌgrædʒʊət‿ˈstjuːdənt]	Student mit Universitätsabschluss
willingness [ˈwɪlɪŋnɪs]	Bereitschaft
safeguard [ˈseɪfgɑːd]	Schutzvorrichtung
cyber crime [ˈsaɪbə ˌkraɪm]	Verbrechen mit Hilfe des Internet
fraudster [ˈfrɔːdstə]	Betrüger(in)
dozen [ˈdʌzn]	Dutzend
chat room [ˈtʃæt‿ˌruːm]	Chatroom (Internet)
thief [θiːf]	Dieb(in)
hacker [hækər]	Hacker, Computereindringling
stranger [ˈstreɪndʒəʳ]	Fremde(r)
proof [pruːf]	Beweis
goodwill [gʊdˈwɪl]	guter Wille
common interest [kɒmən‿ˈɪntrəst]	gemeinsames Interesse

C

retail sales [ˌriːteɪl ˈseɪlz]	Einzelhandelsumsätze

Tapescript

adventurous [ədˈventʃərəs]	abenteuerlich
bay [beɪ]	Ladeplatz, Ladebereich
to break even [breɪk‿ˈiːvn]	kostendeckend arbeiten
carrot [ˈkærət]	Karotte
check-out [ˈtʃekaʊt]	Kasse
chief financial officer [ˌtʃiːf faɪˌnænʃəl ˈɒfɪsəʳ]	Leiter(in) der Finanzabteilung
depot [ˈdepəʊ]	Lagerhaus
grocery [ˈgrɛʊsərɪ]	Lebensmittel-
pallet [ˈpælɪt]	Palette
picker [ˈpɪkəʳ]	hier: Lagerarbeiter, der die bestellte Ware heraussucht
presentation [ˌprezənˈteɪʃn]	Präsentation
queue [kjuː]	Schlange
remote-controlled [rɪˌməʊt‿kənˈtrəʊld]	ferngesteuert
shaky [ˈʃeɪkɪ]	unsicher
stadium [ˈsteɪdɪəm]	Stadion
storey [ˈstɔːrɪ]	Stockwerk, Etage
sweets [swiːts]	Süßigkeiten
third [θɜːd]	Drittel
trolley [ˈtrɒlɪ]	Einkaufswagen

Unit 19

A

fame [feɪm]	Ruhm
extraordinary [ɪkˈstrɔːdɪnrɪ]	außergewöhnlich
businessman [ˈbɪznɪsmæn]	Geschäftsmann
mail order [ˈmeɪlˌɔːdə]	Versand(handel)
based [beɪst]	mit Sitz in
chain [tʃeɪn]	Kette
selection [səˈlekʃn]	Auswahl, Sortiment
Asia [ˈeɪʒə]	Asien
luxurious [lʌgˈʒʊərɪəs]	luxuriös
empire [ˈempaɪə]	Reich, Imperium
to venture; venture [ˈventʃə]	riskieren, sich wagen; Unternehmen
joint venture [ˌdʒɔɪnt ˈventʃə]	Gemeinschaftsunternehmen
to admit [ədˈmɪt]	zulassen, zugeben
human [ˈhjuːmən]	Mensch; menschlich
to motivate [ˈməʊtɪveɪt]	motivieren
cabin crew [ˈkæbɪn ˌkruː]	Besatzung
business-class [ˈbɪznɪs ˌklɑːs]	Businessklasse
branch [brɑːntʃ]	Zweig, Branche
to run the business [ˌrʌn ðə ˈbɪznɪs]	das Geschäft führen

to go ahead (with sth.) [gəʊ‿ə'hed wɪð ˌsʌmθɪŋ]	(etw.) durchführen
record-breaking ['rekɔːd‿ˌbreɪkɪŋ]	Rekord-
amphibious [æm'fɪbɪəs]	Amphibien-; amphibisch
to publicize sth. ['pʌblɪsaɪz‿ˌsʌmθɪŋ]	für etwas Werbung machen
to come up with (an idea) [kʌm‿ˌʌp wɪð‿ən aɪ'dɪə]	(eine Idee) haben, auf etw. kommen
reality show [rɪ'æləti ˌʃəʊ]	Fernsehsendung, die das tägliche Leben normaler Menschen zeigt
ultimately ['ʌltɪmətlɪ]	letztlich, schließlich
affordable [ə'fɔːdəbl]	erschwinglich

B2

total deliveries [ˌtəʊtl dɪ'lɪvərɪz]	Gesamtauslieferungen

Tapescript

airliner ['eəlaɪnə]	Flugzeug
aviation industry [ˌeɪvɪ'eɪʃn ˌɪndəstrɪ]	Luftfahrtindustrie
bold [bəʊld]	kühn
decision-making [dɪ'sɪʒnˌmeɪkɪŋ]	Entscheidungsfindung
domination [ˌdɒmɪ'neɪʃn]	Vorherrschaft
drive [draɪv]	Antrieb, Kampagne
final assembly [ˌfaɪnl ə'semblɪ]	Endmontage
long-range [ˌlɒŋ'reɪndʒ]	Langstrecken-
to put the seal on [ˌpʊt ə 'siːl ɒn]	besiegeln
reputation [ˌrepjə'teɪʃn]	Ruf
rival ['raɪvl]	Konkurrent(in)
to secure [sɪ'kjʊə]	sichern
to simplify ['sɪmplɪfaɪ]	vereinfachen
to streamline ['striːmlaɪn]	rationalisieren
twin-engine [ˌtwɪn'endʒɪn]	mit 2 Motoren
to unify ['uːnɪfaɪ]	vereinheitlichen
unveiling [ʌn'veɪlɪŋ]	Enthüllung
widebody ['waɪdbɒdɪ]	Großraum-

Unit 20

A

globalisation [ˌgləʊbəlaɪ'zeɪʃn]	Globalisierung
controversial [ˌkɒntrə'vɜːʃl]	strittig
harm [hɑːm]	Schaden
World Bank [ˌwɜːld‿'bæŋk]	Weltbank
belief [bɪ'liːf]	Glaube
liberalised ['lɪbrəlaɪzd]	liberalisiert
free market [ˌfriː 'mɑːkɪt]	freier Warenverkehr
desirable [dɪ'zaɪərəbl]	wünschenswert
diet ['daɪət]	Ernährungsweise, Diät
striking ['straɪkɪŋ]	auffallend
impact ['ɪmpækt]	Auswirkung
to render ['rendə]	hier: machen
flu virus ['fluː ˌvaɪərəs]	Grippevirus
to drive [draɪv]	hier: antreiben
distant ['dɪstənt]	entfernt
operation [ˌɒpər'eɪʃn]	Fabrik, Betrieb
integration [ˌɪntɪ'greɪʃn]	Integration
to gain [geɪn]	erhalten, erlangen
to lift [lɪft]	heben, hieven
multi-national corporation [ˌmʌltɪnæʃənl ˌkɔːpə'reɪʃn]	multinationales Unternehmen
advancement [əd'vɑːnsmənt]	Fortschritt
to boost [buːst]	ankurbeln, verstärken
to conduct [kən'dʌkt]	ausführen, abwickeln
disaster [dɪ'zɑːstə]	Katastrophe
concern [kən'sɜːn]	Sorge, Anliegen, Interesse
labour standard ['leɪbə ˌstændəd]	Arbeitsstandard
cost cutting process ['kɒstˌkʌtɪŋ ˌprəʊses]	Kostensenkungsmaßnahme
to violate ['vaɪəleɪt]	verletzen, missachten
to note [nəʊt]	bemerken, erwähnen, notieren
locally ['ləʊkəlɪ]	vor Ort
subsidy ['sʌbsɪdɪ]	Subvention
to perform [pə'fɔːm]	arbeiten, leisten, verrichten
command [kə'mɑːnd]	hier: bieten
commodities [kə'mɒdɪtɪz]	Rohstoffe
raw materials [ˌrɔː mə'tɪərɪəlz]	Rohstoffe
cotton ['kɒtn]	Baumwolle
cocoa ['kəʊkəʊ]	Kakao
capacity [kə'pæsətɪ]	Kapazität, Leistungsvermögen

vulnerable ['vʌlnrəbl]	verwundbar
price swings ['praɪs‿ˌswɪŋz]	Preisschwankungen
shortage ['ʃɔːtɪdʒ]	Mangel
oversupply [ˌəʊvəsə'plaɪ]	Überangebot
vital ['vaɪtl]	entscheidend, lebenswichtig
to struggle ['strʌgl]	sich abmühen
to plummet ['plʌmɪt]	einbrechen, stürzen

B1

fan [fæn]	Ventilator

C1

GDP [ˌdʒiːdiː'piː] (gross domestic product) [ˌgrəʊs dəˌmestɪk 'prɒdʌkt]	BIP (Bruttoinlandsprodukt)
nominal ['nɒmɪnl]	nominal
per capita [pə'kæpɪtə]	pro Kopf

C2

preconditions [ˌpriːkən'dɪʃnz]	Voraussetzungen

Tapescript

allegedly [ə'ledʒɪdlɪ]	angeblich
to demand [dɪ'mɑːnd]	fordern
to dismiss [dɪs'mɪs]	entlassen
gloves [glʌvz]	Handschuhe
hard-pressed [ˌhɑːd'prest]	bedrängt
legal requirement [ˌliːgl rɪ'kwaɪəmənt]	gesetzliche Vorschrift
machinist [mə'ʃiːnɪst]	Maschinenführer(in)
recommendation [ˌrekəmen'deɪʃn]	Empfehlung
retailing ['riːteɪlɪŋ]	Einzelhandel
stuffy ['stʌfɪ]	stickig
union activity [ˌjuːnjən æk'tɪvətɪ]	Gewerkschaftstätigkeit
union leader [ˌjuːnjən 'liːdə]	Gewerkschaftsführer(in)
ventilation [ˌventɪ'leɪʃn]	Belüftung

Unit 21

breakthrough ['breɪkθruː]	Durchbruch
human embryo [ˌhjuːmən 'embrɪəʊ]	menschlicher Embryo
to clone [kləʊn]	klonen
to be at the forefront of sth. [bɪ‿ət‿ðə 'fɔːfrʌnt əv ˌsʌmθɪŋ]	in etw. führend sein
embryonic stem cell [ˌembrɪɒnɪk 'stemˌsel]	embryonale Stammzelle
diabetes [ˌdaɪə'biːtiːz]	Diabetes
go-ahead ['gəʊəhed]	Erlaubnis
to grant [grɑːnt]	bewilligen
licence ['laɪsəns]	Lizenz
authority [ɔː'θɒrɪtɪ]	Behörde
to praise [preɪz]	loben
to extract [ɪk'strækt]	entnehmen
to pave [peɪv]	ebnen
to cure [kjʊə]	heilen
intractable [ɪn'træktəbl]	unlösbar, hartnäckig
medical conditions ['medɪkl kənˌdɪʃnz]	Krankheiten
paralysis [pə'ræləsɪs]	Lähmung
to reignite [ˌriːɪg'naɪt]	wieder entfachen
ethics ['eθɪks]	Ethik
early-stage [ˌɜːlɪ'steɪdʒ]	im frühen Stadium, frühzeitig
critics ['krɪtɪks]	Kritiker
to accuse [ə'kjuːz]	beschuldigen
to trivialise ['trɪvɪəlaɪz]	etw. trivialisieren
handful ['hændfʊl]	handvoll
therapeutic cloning [ˌθerəpjuːtɪk‿'kləʊnɪŋ]	therapeutisches Klonen
ban [bæn]	Verbot
to push for ['pʊʃ fɔː]	auf etw. drängen
to donate [dəʊ'neɪt]	spenden
to undergo [ˌʌndə'gəʊ]	durchmachen, sich unterziehen
in-vitro fertilisation [ɪnˌviːtrəʊ fɜːtɪlaɪ'zeɪʃn]	künstliche Befruchtung
nucleus ['njuːklɪəs]	Zellkern
electrical shock [ɪˌlektrɪkl 'ʃɒk]	elektrischer Schlag
pin [pɪn]	Nadel
blastocyst ['blæstəsɪst]	Blastozyste
beyond [bɪ'jɒnd]	darüber hinaus
genetic defect [dʒəˌnetɪk 'diːfekt]	genetischer Defekt
root [ruːt]	Wurzel
committee [kə'mɪtɪ]	Kommission, Ausschuss
unethical [ˌʌn'eθɪkl]	unethisch

to test [test]	hier: überprüfen
high court [ˌhaɪ 'kɔːt]	oberstes Zivilgericht (GB)
lawyer ['lɔɪə]	Rechtsanwalt/-anwältin
fellowship ['feləʊʃɪp]	Gesellschaft
judicial review [djuːˌdɪʃl 'riːvjuː]	richterliche Überprüfung

Unit 22

peacekeeper ['piːsˌkiːpə]	Blauhelmsoldat (UN)
censure ['senʃə]	Tadel, Kritik
alleged [ə'ledʒd]	vermeintlich, angeblich
slum [slʌm]	Slum
to condemn [kən'dem]	verurteilen, missbilligen
human rights activist [ˌhjuːmən 'raɪts ˌæktɪvɪst]	Menschenrechtsaktivist(in)
cold-blooded [ˌkəʊld'blʌdɪd]	kaltblütig
by contrast [baɪ 'kɒntrɑːst]	im Gegensatz
unsung hero [ˌʌnsʌŋ 'hɪərəʊ]	ein Held, von dem niemand spricht
despot ['despɒt]	Despot
civil strife [ˌsɪvl 'straɪf]	ziviler Unfrieden
anarchy ['ænəkɪ]	Anarchie
performance [pə'fɔːməns]	Leistung
officials [ə'fɪʃlz]	Vertreter
to disagree [ˌdɪsə'griː]	widersprechen
landmark ['lændmɑːk]	hier: Meilenstein
voluntary ['vɒləntrɪ]	freiwillig
repatriation [ˌriːpætrɪ'eɪʃn]	Rückführung
to be scheduled [bɪ 'ʃedjuːld]	geplant sein
humanitarian [hjuːˌmænɪ'teərɪən]	humanitär
to wind up [waɪnd 'ʌp]	enden, beenden
vicious ['vɪʃəs]	teuflisch
civil war [ˌsɪvl 'wɔː]	Bürgerkrieg
to be poised [bɪ 'pɔɪzd]	bereit sein, etw. zu tun
sexual abuse [ˌsekʃʊəl ə'bjuːs]	sexueller Missbrauch
conduct ['kɒndʌkt]	Verhalten
prosecution [ˌprɒsɪ'kjuːʃn]	Strafverfolgung
troops [truːps]	Truppen
to deploy [dɪ'plɔɪ]	einsetzen
personnel [ˌpɜːsən'el]	Personal
civilian staff [sɪˌvɪlɪən 'stɑːf]	zivile Mitarbeiter
incident ['ɪnsɪdənt]	Vorfall
civilian casualties [sɪˌvɪlɪən 'kæʒjʊəltɪz]	zivile Opfer
to fail [feɪl]	scheitern, missglücken, durchfallen
popularly elected [ˌpɒpjʊləlɪ‿ɪ'lektɪd]	allgemein gewählt
to abide [ə'baɪd]	befolgen
interim ['ɪntərɪm]	vorläufig, provisorisch
verdict ['vɜːdɪkt]	Urteil
legitimate [lə'dʒɪtɪmət]	legitim
to support [sə'pɔːt]	unterstützen
in accordance with [ɪn ə'kɔːdəns wɪð]	in Übereinstimmung mit
observer [əb'zɜːvə]	Beobachter(in)
related [rɪ'leɪtɪd]	bezogen, verwandt
misbehaviour [ˌmɪsbɪ'heɪvjə]	Fehlverhalten
impartial [ɪm'pɑːʃl]	unparteiisch
failure ['feɪljə]	Versagen,
misconduct [ˌmɪs'kɒndʌkt]	Fehlverhalten
to prosecute ['prɒsɪkjuːt]	strafrechtlich verfolgen

Unit 23

to call for ['kɔːl fɔː]	verlangen nach
junk food ['dʒʌŋk ˌfuːd]	(ungesunde) Schnellgerichte
to reverse [rɪ'vɜːs]	hier: ändern (Gewohnheit)
eating habits ['iːtɪŋ ˌhæbɪts]	Essgewohnheit
obesity [ə'biːsətɪ]	Fettleibigkeit
to launch [lɔːntʃ]	hier: starten, beginnen
blueprint ['bluːprɪnt]	hier: Plan, Programm
pernicious [pə'nɪʃəs]	schädlich
overwhelming [ˌəʊvə'welmɪŋ]	überwältigend
rampant ['ræmpənt]	ungezügelt
nutrition [njuː'trɪʃn]	Ernährung
to target ['tɑːgɪt]	auf jmdn. abzielen (Werbung)
to compensate for ['kɒmpənseɪt fɔː]	etwas ausgleichen, wettmachen
deficiency [dɪ'fɪʃənsɪ]	Defizit
to opt out of [ɒpt‿'aʊt‿əv]	aus etwas aussteigen
row [raʊ]	Streit, Krach
voucher ['vaʊtʃə]	Gutschein
reward [rɪ'wɔːd]	Belohnung
attendance [ə'tendəns]	Teilnahme
wrapper ['ræpə]	Verpackung
four-pronged [ˌfɔː'prɒŋd]	Vier-Punkte-
to shape [ʃeɪp]	bilden, formen
well-placed [ˌwel'pleɪst]	gut platziert

Unit 24

multiculturalism [ˌmʌltɪˈkʌltʃərəlɪzm]	Multikultur
separateness [ˈsepərətnəs]	Distanz
to rediscover [ˌriːdɪˈskʌvə]	wieder entdecken
core [kɔː]	Kern
mosque [mɒsk]	Moschee
by the same token [baɪ ðə ˌseɪm ˈtəʊkn]	aus demselben Grund
dual [ˈdjuːəl]	Doppel-, zweierlei
heritage [ˈherətɪdʒ]	Erbe
for goodness sake [fə ˌgʊdnɪs‿ˈseɪk]	Du liebe Güte!
to spur [spɜː]	anspornen
to cease [siːs]	aufhören, (be)enden
objective [əbˈdʒektɪv]	Ziel
mainstream [ˈmeɪnstriːm]	Hauptrichtung, Mainstream
figurehead [ˈfɪgəhed]	Galionsfigur

Skills files

to skim [skɪm]	(Text) überfliegen
to scan [skæn]	(Text nach bestimmten Details) durchsuchen
subtitle [ˈsʌbˌtaɪtl]	Untertitel
subheading [ˈsʌbˌhedɪŋ]	Zwischenüberschrift
image [ˈɪmɪdʒ]	Abbildung
word-by-word [ˌwɜːd‿baɪ ˈwɜːd]	Wort für Wort
margin [ˈmɑːdʒɪn]	Rand, Seitenrand
skilled [skɪld]	geschult

AE = American English
BE = British English
T = Tapescript

A

a bit 16 ein bisschen
to a great extent 18 in großem Umfang
A level 98 etwa: Abitur
to abide 140 befolgen
to abolish 118 abschaffen
aborigine 65 austral. Ureinwohner(in)
abortion 116, T Abtreibung
above 11 oben
absolute majority 72, T absolute Mehrheit
to abuse; abuse 114 missbrauchen; Missbrauch
accent 43 Akzent
acceptable 118 akzeptabel, annehmbar
access 35 Zugang
to accomodate 110 unterbringen
accomodation 71 Unterkunft
according to 39 (seiner/ihrer) Meinung nach, laut
accountancy 10 Rechnungswesen
accounts department 94 Buchhaltung
accusation 74 Anschuldigung
to accuse 138 anklagen, beschuldigen
achievement 101 Erfolg
acid rain 50 Saurer Regen
(well-) acquainted 74 (gut) bekannt
active 17 aktiv
actually 112 tatsächlich
to adapt 18 bearbeiten
addict 74 Abhängige(r)
addiction 34 Sucht, Abhängigkeit
additional 43 zusätzlich
add-ons 98 Zusatzdienste
to address 99 adressieren an
to address people 97 jmdn. anreden
adequate 82 ausreichend
adjective 18 Adjektiv
to admit 130 zulassen, zugeben
advanced 125 fortschrittlich
advancement 134 Fortschritt
adventure 14 Abenteuer
adventurous 128, T abenteuerlich
advert 40 Kurzform von "advertisement"= Werbeanzeige
advertiser 39 Inserent(in), Auftraggeber(in) von Werbesendungen
advertising campaign 50 Werbekampagne
advertising manager 17 Werbeleiter(in)
advice 72 Rat
to advise 32 beraten
to affect 34 beeinflussen, sich auswirken auf
affection 91 Zuneigung, Liebe
to afford 19 sich leisten
affordable 132 erschwinglich
aged ... 24 Jahre alt
agenda 111 Tagesordnung, Programm
aid 43 Hilfe, Hilfsmittel
to aim (at) 40 zielen (auf)
air-conditioner 122 Klimaanlage
airliner 132,T Flugzeug
airtime 39 Sendezeit
aisle seat 108 Gangplatz
alcohol 118 Alkohol
alcoholic 114 Alkoholiker(in)
alive 18 lebend, lebendig, aktiv
all over the world 51 in der ganzen Welt
alleged 140 vermeintlich, angeblich
allegedly 136, T angeblich
to ally; ally 122 verbünden; Verbündete(r)
American Indian 84 amerikanischer Indianer, amerikanische Indianerin
amnesia 118 Amnesie, Gedächtnisverlust
amount 22 Betrag, Menge
amphibious 131 Amphibien-; amphibisch
analyst 126 Analyst(in)
anarchy 140 Anarchie
angel 73 Engel
antonym 58 Antonym, Wort mit gegensätzlicher Bedeutung
(not) any more 54 nicht länger
apparent 91 offensichtlich
appearance 101 Erscheinungsbild
apple pie 95 Art Apfelstrudel
applicant 24 Bewerber(in)
application 10 Anwendung; Bewerbung
to apply 24 sich bewerben
to appoint 72 ernennen
appointment 104 Termin
apprenticeship 98 Ausbildung
approach 54 Ansatz, Annäherung
approach (to business) 96 Herangehensweise bei einem Geschäft
appropriate 75 angemessen, angebracht
to arrange 104 ausmachen, arrangieren, zuwege bringen
arrival 69 Ankunft
as a result 62 folglich
as well as 14 sowie
Asia 130 Asien
assistant 24 Assistent(in)
to associate 40 assoziieren, in Verbindung bringen
astounded 92 verblüfft, überrascht
at all costs 50 um jeden Preis
at the latest 108 spätestens
atmosphere 122 Atmosphäre
to attach 110 anhängen
attempt 122 Versuch
to attend 91 teilnehmen
attendance 142 Teilnahme
attitude 22 Haltung, Einstellung
attraction 56 Anziehungskraft
attractive 46 reizvoll, anziehend, attraktiv
auction 126 Auktion
audio 42 audio, Hör-
authority 138 Behörde
Automobile Sales Management Assistant 99 Automobilkaufmann, -frau
availability 58 Verfügbarkeit
available 43 verfügbar, vorhanden
average 22 Durchschnitt
aviation industry 132,T Luftfahrtindustrie

to avoid 43 vermeiden
avoidance 125 Vermeidung
aware 100 bewusst
awareness 124,T Erkenntnis
awesome 69 großartig

B

background 39 hier: Hintergrund, Herkunft, Verhältnisse
backpacking 66 Rucksackreisen
baggage 92 Gepäck (AE)
Baltic 30 baltisch; Ostsee
ban 138 Verbieten; Verbot
bandanas 91 Halstuch, Kopftuch
Bank Business Management Assistant 99 Bankkaufmann, -frau
Bank Holiday 22 (öffentlicher) Feiertag
baptist 91 Baptist(in)
bar graph 35 Säulendiagramm
bargain 126 günstiges Angebot, Sonderangebot
barrel 64, T Fass
base 48 Basis, Stützpunkt, Ausgangspunkt
based 130 mit Sitz in
basics 69 Grundlagen
bather 122 Badende(r)
bathroom 91 Badezimmer, Toilette
bay 128, T Ladeplatz, Ladebereich
to be at the forefront of sth. 138 in etw. führend sein
to be committed to doing sth 32 sich dafür einsetzen, etw. zu tun
to be elected 72,T gewählt werden
to be engaged 106 verplant sein
to be fortunate 72,T Glück haben
to be into 14 Fan von etwas sein
to be involved in 103 an etwas beteiligt sein
to be poised 141 bereit sein, etw. zu tun
to be scheduled 141 geplant sein
to be unfamiliar with sth. 74 etw. nicht kennen, mit etw. nicht vertraut sein
beat 78 hier: Takt
bedsitter 11 Einzimmerwohnung
beforehand 101 zuvor
to beg 116,T betteln
beginner 11 Anfänger(in)
to behave 110 sich benehmen
behaviour 90 Verhalten, Benehmen
Belgium 27 Belgien
belief 134 Glaube
benefit 22 Zuwendung, soziale Leistung
to benefit 103 profitieren
best known 39 bekannteste(r, s)
beyond 138 darüber hinaus
billion 58 Milliarde
bills 72 hier: Gesetzentwurf
binge drinking 116,T Besäufnis, Saufgelage
birth control 114 Geburtenkontrolle
biscuit 14 Keks
blastocyst 138 Blastozyste
to blow one's nose 91 sich die Nase putzen
blueprint 142 hier: Plan, Programm
bold 132,T kühn
bold type 15 Fettdruck, fett gedruckt
to boost 134 ankurbeln, verstärken
boundary 30 Grenze
branch 130 Zweig, Branche
brand name 40 Markenname
to break even 128,T kostendeckend arbeiten
breakthrough 138 Durchbruch
briefly 98 kurz
bright 112 hell
broadband 35 Breitband
broadcaster 34 Rundfunksender
to broaden the mind 14 den Horizont erweitern
brochure 70 Broschüre
BTEC National Diploma 9 etwa: Fachhochschulreifeprüfung
BTU British Thermal Unit 59 BTU = brit. Wärmeeinheit
to build up 103 aufbauen
bulb 50 (Glüh)birne
bumpy 112 holprig, unruhig
bureaucracy 26 Bürokratie
bus pass 10 Busausweis
business associate 109 Geschäftspartner
business protocol 97 Business-Etikette
business studies 9 etwa: Fachbereich Wirtschaft
business-class 130 Businessklasse
businessman 130 Geschäftsmann
button 42 Knopf
by contrast 140 im Gegensatz
by rail 48 mit der Bahn
by the same token 144 aus demselben Grund

C

cabin crew 130 Besatzung
cable 62 Kabel
CAD (Computer Aided Design) 10 Computerunterstützte Zeichnung
to call at 48,T anhalten
to call for 142 verlangen nach
to calm down 116,T beruhigen
CAM (Computer Aided Manufacturing) 10 Computerunterstützte Fertigung
Cambodia 88,T Kambodscha
campaign 39 Kampagne, Werbefeldzug
campus 11 Campus, Universitätsgelände
can 91 Dose
can't stand 15 nicht ausstehen können
to cancel 106 absagen
candidate 24 Kandidat(in), Bewerber(in)
canteen 113 Kantine
capacity 135 Kapazität, Leistungsvermögen
capital city 25 Hauptstadt
to capitalize 110 groß schreiben
caption 70 Titel, Überschrift
car hire 98 Autovermietung
carbon paper 124,T Kohlepapier
card 55 Karte
cardboard packaging 125 Kartonverpackung
to care for 24 sich um etw. kümmern
carrot 128,T Karotte
to carry out 23 ausführen
carry-on-luggage 66 Handgepäck
cartoon figures 68 Karikaturen
cartoonist 70 Karikaturist(in)
case study 121 Fallstudie
cashier 91 Kassierer(in)
cattle 69 Vieh
to cause 78 verursachen
to cease 145 aufhören, (be)enden
cell 62 Zelle
censure 140 Tadel, Kritik

centigrade 122 Celsius
certificate 9 Zeugnis
CFC 51 FCKW (Fluorchlorkohlenwasserstoff)
chain 130 Kette
chain letter 110 Kettenbrief
chainsaw 69 Kettensäge
challenge 24,T Herausforderung
challenging 98 herausfordernd
chancellor 65 Kanzler(in)
change 114 hier: Wechselgeld
charity 114 karitative Organisation, Wohlfahrt
chart 49 Tabelle, Schaubild
chartered flight 98 Charterflug
chat room 126 Chatroom (Internet)
to check 104 überprüfen
check-out 128,T Kasse
chemical 122 chemisch
chemist 51 Chemiker(in)
Chief Accountant 113 Hauptbuchhalter(in)
chief financial officer 128,T Leiter(in) der Finanzabteilung
childcare assistant 24 Kinderpfleger(in)
Chinese 84 chinesisch
chlorofluorocarbon (CFC) 122 Fluorchlorkohlen-wasserstoff (FCKW)
church-run 114 kirchlich
circle 86 Kreis, Kreislauf
citizen 25 Bürger(in)
civil engineering 10 Bauwesen
civil strife 140 ziviler Unfrieden
civil war 140 Bürgerkrieg
civilian casualties 140 zivile Opfer
civilian staff 140 zivile Mitarbeiter
to claim 34 behaupten, verlangen, beanspruchen, Anspruch geltend machen
client 105 Kunde, Kundin
climate 88 Klima
"clock conscious" 97 hier: mit Sinn für Pünktlichkeit
to clone 138 klonen
to close a deal 96 Geschäft abschließen
to close down 19 schließen, zumachen (z. B. Firma)
closet 92 Schrank (AE)
clothing 117 Kleidung
CO2 122 Kohlendioxyd
coach 11 Trainer(in)
coaching days 48 Postkutschenzeit
coastal 56 Küsten-
cocoa 135 Kakao
cod 95 Kabeljau
to code; code 54 kodieren; Kode
cognizant 124,T bewusst
cold 91 Erkältung
cold-blooded 140 kaltblütig
column 27 Spalte
to come into operation 27 in Kraft treten
to come to hand 114 zur Verfügung stehen
to come true 54 wahr werden
to come up with (an idea) 131 (eine Idee) haben, auf etw. kommen
command 43 Befehl, Führung
command 98 hier: Sprachbeherrschung
to command 135 hier: bieten
comment 14 Bemerkung, Kommentar
commercial 39 Werbespot
commercial 56,T gewerblich, kommerziell
to commit 101 hier: investieren
committee 138 Kommission, Ausschuss
commodities 135 Rohstoffe
common 32 gemeinsam
common interest 126 gemeinsames Interesse
communication skills 101 Kommunikationsfähigkeit
communications 33 Telekommunikation, Kommunikationswege
communist era 30 kommunistisches Zeitalter
community 27 Gemeinschaft, Bevölkerungsgruppe, Gemeinde
commuter 46 Pendler(in)
Company Training Manager 113 Ausbildungsleiter(in)
to compensate for 142 etwas ausgleichen
competitor 32,T Wettbewerber(in)
complacent 124,T selbstzufrieden
complaint 105 Beschwerde
composting 49 Kompostierung
comprehensive 98 umfassend
comprehensive school 9 Gesamtschule
compulsory 72,T verpflichtend
concern 135 Sorge, Anliegen, Interesse
concerned 56 betroffen, interessiert
to conclude 71 folgern
to condemn 140 verurteilen, missbilligen
condition 22 Bedingung
to conduct 134 ausführen, abwickeln
conduct 140 Verhalten
conference room 112 Konferenzzimmer
confidence 97 Zuversicht, Vertrauen
confident 40,T hier: sicher
confident 111 zuversichtlich
to confirm 86 bestätigen
to confront 122 gegenüberstellen, konfrontieren
confused 118 verwirrt
congested 42 überfüllt, verstopft
congestion 42 Stau
connotation 97 Bedeutung
conscious 50 bewusst
consecutive 82 folgend
consensus view 82 übereinstimmende Ansicht
consequence 122 Folge, Konsequenz
conservative 72 Konservative(r)
to consist of 32 bestehen aus
constant 59 ständig, gleichbleibend
constituency 72 Wahlkreis
construction 34 Bau, Bauweise, Konstruktion
consumerism 124,T Konsumdenken
consumption 50 Verbrauch
Continent 98 Kontinent
contrary to 91 im Gegensatz zu
to contribute 50 beitragen
contribution 50 Beitrag
control 10 Kontrolle, hier: Regeltechnik
controversial 134 strittig
convenient 46 günstig, praktisch, handlich
to converse 97 sich unterhalten
to convey 97 übermitteln
cookies 92 Keks, Plätzchen
to cope (with) 14 mit etw. fertig werden

core 144 Kern
corn 86 Getreide, Mais, Korn
corrections 68 Korrekturen
correspondence 111 Korrespondenz, Briefverkehr
cosmopolitan 30 kosmopolitisch
cost cutting process 135 Kostensenkungsmaßnahme
cotton 135 Baumwolle
country code 104,T Landesvorwahl
countryside 30 Land, Landschaft
cover 66 Deckung (Versicherung)
covering letter (= letter of application) 98 Anschreiben
crack addict 74 Crack-Abhängige(r)
to crawl 73 kriechen
crazy 46 verrückt
to create 56 schaffen, verursachen
creative 17 kreativ
crime 118 Verbrechen
crime level 118 Verbrechensrate
crisis 126 Krise
critics 138 Kritiker
cross-cultural 94 interkulturell
crossroads 48 Kreuzung
crucial 97 entscheidend, äußerst wichtig
crude oil 64,T Rohöl
cultural 117 kulturell
to cure 138 heilen
currency 27 Währung
custard 95 Vanille(soße)
customer service 98 Kundenservice
to cut 122 reduzieren
CV = curriculum vitae 24 Lebenslauf
cyber crime 126 Verbrechen mit Hilfe des Internet

D

damp 117 feucht
dancer 17 Tänzer(in)
Data Processing 113 Datenverarbeitung
day by day 42 Tag für Tag
dealer 54 Händler(in)
Dear Madam 24 Sehr geehrte Dame
Dear Sir 24 Sehr geehrter Herr
Dear Sir or Madam 99 Sehr geehrte Damen und Herren
decade 82 Jahrzehnt
decision-making 132,T Entscheidungsfindung
to declare 114 verkünden
decline 64,T Rückgang
to decrease 39 abnehmen, vermindern
deficiency 142 Defizit
definite 47 eindeutig, bestimmt
degree 18 akad. Grad (z.B. B.A., M.A.)
dehydration 78 Austrocknung, Dehydration
delay 48 Verspätung
delicate 124,T heikel
delighted 111 erfreut
demand 64,T Nachfrage
to demand 136 fordern
democrat 72 Demokrat(in)
democratic 27 demokratisch
Denmark 27 Dänemark
departure 108 Abflug, Abreise
dependent 115 abhängig, hier: minderjährig
to deploy 140 einsetzen
deposit 56 Lagerstätte, Guthaben
depot 128,T Lagerhaus
deprivation 114 Entbehrung, Entzug
deprived 118 unterprivilegiert
desirable 134 wünschenswert
to desire 101 wünschen
desk 108 hier: Schalter
despite 82 trotz
despot 140 Despot
dessert 114 Nachtisch
destination 42 Reiseziel, Zielort
detour route 48 Umleitung
developing country 58 Entwicklungsland, Industrieland
device 33 Gerät
devil 73 Teufel
diabetes 138 Diabetes
dialogue 11 Dialog
diary 107 Terminplan
diet 134 Ernährungsweise, Diät
to dig up 62 (aus)graben
dilemma 82 Klemme, Dilemma
dinosaur 122 Dinosaurier
diplomas 101 Diplome, Prüfungsurkunden
to direct 113 leiten
direct speech 51 direkte Rede
disabled 109 behindert
to disagree 140 widersprechen
disagreement 51 abweichende Meinung, Meinungsverschiedenheit
disappearance 30 Verschwinden
disaster 134 Katastrophe
disciplined 91 diszipliniert
discotheque 116 Disko
discount 26 Rabatt
to discover 17 entdecken
to disguise 80,T verschleiern, tarnen
to dismantle 54 zerlegen, demontieren
to dismiss 136 entlassen
display 42 Anzeige
to display 111 zeigen, anzeigen
disposal 50 Beseitigung, Entsorgung
to dispose of 124,T entsorgen
distant 134 entfernt
distinction 79 Unterscheidung
to distinguish 126 unterscheiden
distribution 72 Verteilung
district 14 Gegend
to dive 67 tauchen
diversion 43 Umleitung
diversion sign 48,T Umleitungsschild
diversity 30 Vielfalt, Verschiedenheit
divorced 99 geschieden
document shred 124 Papierwolle
domestic 62 häuslich
domestic trip 31 Inlandsausflug
to dominate 34 dominieren, beherrschen
domination 132,T Vorherrschaft
to donate 138 spenden
dot.com 126 Dotcom- (Firma), Online-(Geschäfte)
double bed 110 Doppelbett
double room 109 Doppelzimmer
doubtless 111 zweifellos
dozen 126 Dutzend
to draw up 110 aufstellen
drive 132,T Antrieb, Kampagne
to drive 134 hier: antreiben
to drop 59 fallen, fallen lassen
drug 118 Droge
drug counsellor 78 Drogenberater (in)
dual 144 Doppel-, zweierlei
due to 43 aufgrund von
to dump; dump 54 (Müll) abladen; Müllkippe
Dutch 22 niederländisch; Niederländer(in)
dynamic 103 dynamisch

E

early-stage 138 im frühen Stadium, frühzeitig
earnings 86 Einkommen, Verdienst, Ertrag
eastbound section 48 (Fahrbahn in) östlicher Richtung
eating habits 142 Essgewohnheiten
ecological 46 ökologisch, umweltfreundlich
ecology 56 Ökologie
economics 26 Wirtschaft-(swissenschaft), Volkswirtschaft, Volkswirtschaftslehre
economist 86 Wirtschaftswissen-schaftler(in)
Ecstasy 78 Ecstasy (Name einer Droge)
editor 88 Herausgeber(in), Redakteur(in)
to elaborate 101 näher ausführen
election 27 Wahl
electrical engineering 10 Elektro-technik
electrical shock 138 elektrischer Schlag
electrically-driven 46 elektrisch angetrieben
electronics factory 16,T Elektronik-fabrik
to eliminate 72,T aussondern
elsewhere 51 woanders
embassy 66 Botschaft
embryonic stem cell 138 embryonale Stammzelle
emission 50 Emission, Ausstoß
emphasis 51 Betonung
to emphasize 50 betonen
empire 130 Reich, Imperium
employee 22 Arbeitnehmer(in), Angestellte(r)
to enable 100 ermöglichen
to enclose 24 beilegen (als Anlage)
encouragement 117 Ermutigung
engineering studies 9 Ingenieur-wissenschaft
English Channel 122 Ärmelkanal
enormous 18 gewaltig, enorm
to enrol 100 sich immatrikulieren
to enter 43 eingeben (Daten)
entertainment 71 Unterhaltung
enthusiastic 16 begeistert
entrance exam 26 Aufnahmeprüfung
environmental 50 Umwelt-
environmentalist 46 Umwelt-schützer(in)
environmentally-friendly 57 umweltfreundlich
enzymes 124,T Enzyme
to equip 42 ausrüsten, ausstatten mit
equipment 10 Ausrüstung, Ausstattung
equivalent 98 entsprechend
essay 34 Abhandlung, Aufsatz
essential 24 wesentlich, unentbehrlich
to establish 32 einrichten, gründen
to estimate 46 schätzen
Estonia 30 Estland
ethics 138 Ethik
etiquette 96 Etikette
European Commission 25 Europä-ische Kommission
European Union 25 Europäische Union (die EC wurde im November 1993 in die EU umbenannt)
to evaluate 101 beurteilen
event 34 Veranstaltung
evolution 122 Entwicklung, Evolution
exaggeration 122 Übertreibung
examination 100 Prüfung
to exceed 124,T übersteigen
except 22 außer
exchange 14 Austausch
exchange year 32 Austauschjahr
to expand 18 ausdehnen, ausbreiten
expectation 114 Erwartung
to expend 124,T aufwenden
expense 71 Ausgaben
expert 46 Experte, Expertin
exploration block 64,T Erkundungs-plattformen
export assistant 24 etwa: Außenhandelskaufmann, -frau
Export Manager 113 Abteilungs-leiter(in) Export
to expose 91 entblößen
extension 104,T Durchwahl
extinction 122 Aussterben
to extract 138 entnehmen
extraordinary 130 außergewöhnlich
eye contact 96 Blickkontakt

F

to face sth. 46 mit etw. konfrontiert sein
facilities 109 Einrichtungen
factor 27 Faktor
factual 111 sachlich
to fail 140 scheitern, missglücken, durchfallen
failure 141 Versagen,
fall 92 Herbst (AE)
fame 130 Ruhm
family-sized 40,T für Familien zugeschnitten
fan 136 Ventilator
Far East 103 Ferner Osten
far-reaching 82 weitreichend
fast running 64 schnell fließend
fault 116 Schuld
fear 56 Furcht
feature 43 Merkmal, Bestandteil
fellowship 138 Gesellschaft
female 11 weiblich
fencing 69 Einzäunung
to fetch 114 holen, abholen
figure 35 Zahl, Figur
figurehead 145 Galionsfigur
to file 32,T abheften, ablegen
to fill in 26 ausfüllen
final assembly 132,T Endmontage
finally 99 schließlich
finance 10 Finanzwesen
Finance and Personnel Department 113 Finanz- und Personalabteilung
fireworks displays 69 Feuerwerke
first floor 113 Erdgeschoss (AE), 1. Obergeschoss (BE)
fixed-route 46 schienengebunden
flame 122 Flamme
flashy 101 auffallend
flexible 46 flexibel
flight 74 Flucht
flight attendant 17 Flugbegleiter(in)
flight schedule 108 Flugplan
to flood 122 überfluten
floorplan 113 Grundriss
flu virus 134 Grippevirus
fluency 98 Sprachbeherrschung
foam 122 Schaum

fog 43 Nebel
food stamp 86 Essensmarke
food wrappers 124 Verpackung von Nahrungsmitteln
food-processing company 88,T Nahrungsmittelunternehmen
for better or for worse 122 zum Besseren oder Schlechteren hin
for goodness sake 144 Du liebe Güte!
for instance 16 beispielsweise
forbidden 110 verboten
forest 122 Wald
forgetfulness 118 Vergesslichkeit
form of address 99 Anrede
former 72 frühere(r,s), erstgenannte(r,s)
formula 11 Formel
fortune 18 Glück, Reichtum
fossil 57 fossil
fossil fuels 58 fossile Brennstoffe
foster care 74 bei Pflegeeltern
to found 69 gründen
foundation 97 Stiftung
four-pronged 142 Vier-Punkte-
fraudster 126 Betrüger(in)
free market 134 freier Warenverkehr
freedom 46 Freiheit
freeway 92 Autobahn (AE)
French fries 92 Pommes frites (AE)
to freshen up 94 sich frisch machen
Friends of the Earth 54 Name einer Umweltschutzorganisation
frowned upon 99 verpönt
fuel 51 Brennstoff, Kraftstoff
fuel cell 62 Brennstoffzelle
full board 109 Vollpension
full-time 10 Vollzeit
fumes 42 Rauch, Dämpfe, Abgase
furthermore 51 außerdem, ferner

G

to gain 134 erhalten, erlangen
gambler 114 Spieler(in)
gang 117 Bande
gap 97 Lücke
gap year 32 etwa: Lückenjahr
garbage 122 Abfall, Müll
gas 92 Benzin (AE), Gas
to gather 32,T sammeln
GCSE = General Certificate of Secondary Education 9 etwa: Fachoberschulreife
GDP (gross domestic product) 137 BIP (Bruttoinlandsprodukt)
gender 97 Geschlecht
general science 10 allgemeine Naturwissenschaften
to generate 62 erzeugen
genetic defect 138 genetischer Defekt
genius 114 Genie
Gents 113 Herren (WC)
geographical 58 geografisch
Germany 27 Deutschland
to get across 34 verständlich machen
to get along 91 gut miteinander auskommen
to get on like a house on fire 14 sich sehr gut verstehen
to get on with 22 auskommen mit
to get rid of 46 loswerden, abschaffen
to get to know 14 kennenlernen
to get used to 42 sich gewöhnen an
gift 97 Geschenk
glamorous 78 schick, glamourös
glance 43 Blick
glass-walled 30 mit Wänden aus Glas
global warming 58 weltweiter Temperaturanstieg
globalisation 134 Globalisierung
gloves 136 Handschuhe
to go ahead (with sth.) 130 (etw.) durchführen
to go missing 74 verschwinden
go-ahead 138 Erlaubnis
goodwill 126 guter Wille
government regulation 124 Vorschriften der Regierung
governmental agencies 124,T Regierungsstellen
Governor-General 72 Generalgouverneur
GPS (Global Positioning System) 42 GPS-Navigationssystem
grade 91 Abschluss, Note
gradual 59 allmählich, sanft (ansteigen/abfallen)
gradually 30 allmählich
to graduate 18 Schulabschluss machen, Schule oder Universität beenden, einen akad. Grad erwerben
graduate student 126 Student(in) mit Universitätsabschluss
grammar school 102 Gymnasium
grant 10 Zuschuss, Bafög
to grant 138 bewilligen
graph 35 Diagramm, Schaubild, mathematische Kurve
to grasp 124,T begreifen
grateful 111 dankbar
Great Britain 27 Großbritannien
Greece 27 Griechenland
greenhouse effect 122 Treibhauseffekt, globale Erwärmung
grid 62 Versorgungsnetz
grocery 128,T Lebensmittel-
ground floor (BE) 113 Erdgeschoss
to guide 42 leiten, führen

H

habit 39 Gewohnheit
hacker 126 Hacker, Computereindringling
hairdo 101 Frisur
hairdresser 17 Friseur(in)
half board 109 Halbpension
hall 91 Flur
to halt; halt 122 anhalten; Halt, Pause, Unterbrechung
handful 138 handvoll
hard-pressed 136 bedrängt
hard-working 16 fleißig
harm 134 Schaden
to harm 58 schädigen
harmful 122 schädlich
to have in common 22 gemeinsam haben
head of department 14 hier: Fachbereichsleiter(in)
heading 46 Überschrift
head-on 83 Frontal-, frontal
headquarters 103 Hauptquartier
healing 73 Heilung
health care 10 Gesundheitsfürsorge
heavy metal 14 Heavy Metal (harte Rockmusik)
Here you go. 114 Bitte schön.
heritage 144 Erbe
high court 138 oberstes Zivilgericht (BE)

higher secondary modern school 102 Realschule
higher secondary vocational school 102 Fachoberschule, Berufskolleg
high-speed 30 Hochgeschwindigkeits-
to hike 67 wandern
hiring decision 99 Einstellungsentscheidung
hold-up 48,T Stau
holidaymaker 31 Urlauber(in)
holidaymaking 31 in den Urlaub fahren
Holy Bible 122 (Heilige) Bibel
homeless 74 obdachlos
(to be) homesick 88,T Heimweh haben
hooligan 118 Hooligan, Randalierer
hopelessness 78 Hoffnungslosigkeit
host 14 Gastgeber(in)
hostel 69 Herberge
hotel owner 56,T Hoteleigentümer(in)
household 35 Haushalt
household goods 103 Haushaltswaren
housing estate 56 Siedlung
to hug 96 in den Arm nehmen
huge 30 riesig
human 130 Mensch; menschlich
human embryo 138 menschlicher Embryo
human rights activist 140 Menschenrechtsaktivist(in)
humanitarian 140 humanitär
humanity 122 Menschheit, Menschlichkeit
humour 110 Humor
to hunt 126 jagen
hunter-gatherer 69 Jäger und Sammler
hydro 62 Wasser-

I

to identify 22 identifizieren
illusion 114 Illusion, Einbildung
image 146 Abbildung
immigrant 81 Einwanderer(in)
immigration 81 Einwanderung
immoral 118 unmoralisch
immune system 122 Immunsystem
to impact 97 beeinflussen
impact 134 Auswirkung
impartial 141 unparteiisch
to impede 124,T verhindern, behindern
impersonal 99 unpersönlich
impolite 91 unhöflich
impression 58 Eindruck
impressive 101 beeindruckend
improvement 51 Verbesserung
in accordance with 141 in Übereinstimmung mit
in care 74 in Pflege
in conclusion 50 abschließend
in favour of 47 zu Gunsten von
in general 22 im Allgemeinen
in my view 16 meiner Ansicht nach
in the long run 86 auf lange Sicht
in the meantime 18 in der Zwischenzeit
in the mid 62 in der Mitte von, Mitte der ...
inability 124,T Unvermögen
inadequate 117 nicht ausreichend, unangemessen
incident 140 Vorfall
incineration 49 Verbrennung
income 19 Einkommen
income support 118 Sozialunterstützung
to incur 124,T übernehmen
Independence Day 93 Unabhängigkeitstag
independent 17 unabhängig
indicator 118 Anzeiger, Indikator
individual 46 individuell
to individualise 99 individualisieren
individually 31 einzeln
individuals 98 Personen
industrial 10 industriell
industrial region 103 Industrieregion
industrialized 58 industrialisiert
infertile 122 unfruchtbar
inflow 86 Zustrom
to influence 50 beeinflussen; Einfluss
informal 15 informell
information sheet 11 Informationsblatt
ingredients 78 Bestandteile
initial 98 anfänglich
ink 124,T Tinte
inner 46 Innen-, inner-
innovation 42 Neuerung
to insist (on) 114 beharren, bestehen (auf)
to install 43 einbauen
institution 32 Einrichtung
instrumentation 10 Messtechnik
to insulate 50 isolieren
insulation 51 Isolierung
insult 97 Beleidigung
insurance 98 Versicherung
insurance fraud 121 Versicherungsbetrug
to insure 26 versichern
integral part 97 integraler Bestandteil
integration 134 Integration
intention 32 Absicht
interaction 126 Interaktion
interactivity 68 Interaktivität
interest rates 124 Zinsen
interim 140 vorläufig, provisorisch
intern 100 Praktikant(in)
internet address 68 Internetadresse
internship 32 Praktikum
interval 43 zeitlicher Abstand, Intervall
intractable 138 unlösbar, hartnäckig
in-vitro fertilisation 138 künstliche Befruchtung
to involve 24,T einschließen
Ireland 27 Irland
irreplaceable 122 unersetzlich
issue 56 Thema, Frage
Italy 27 Italien
item 111 Punkt, Posten, Gegenstand, Artikel

J

jew 88,T Jude
jewellery 101 Schmuck
to join 62 sich anschließen
joint 14 gemeinsam
joint venture 130 Gemeinschaftsunternehmen
joyride 121 (wilde) Vergnügungsfahrt
judge 118 Richter(in)
judicial review 138 richterliche Überprüfung
jumper 66 Pullover
junk 114 Gerümpel, Ramsch

junk food 142 (ungesunde) Schnellgerichte
junkie 34 Abhängige(r)

K

kangaroo 70 Känguru
to keep up with 74 Schritt halten mit
Khmer Rouge 88,T Rote Khmer
kid 114 Kind, Jugendlicher
kindness 114 Güte, Freundlichkeit
kitchen equipment 104,T Küchengeräte
to know by name 26 mit Namen kennen

L

to label 66 beschriften
Labor Day 93 Tag der Arbeit
labor pool 82 Arbeitsmarkt
labour 72 Arbeit
labour standard 135 Arbeitsstandard
lack 122 Mangel
to lag 124,T hinterherhinken
land bridge 69 Landbrücke
landfill 49 Mülldeponie
landfill deposit 125 Mülldeponie
landmark 140 hier: Meilenstein
to launch 34 etw. starten, auf den Markt bringen, einführen
to launch 142 hier: starten, beginnen
lavish (gifts) 97 üppig, großzügig
lawyer 138 Rechtsanwalt, Rechtsanwältin
layabout 79 Faulenzer(in)
layout 68 Gestaltung
leadership 30 Führung, Leitung
leading 24 führend
least 31 am wenigsten
lecture 26 Vorlesung
left 54 übrig
legal requirement 136 gesetzliche Vorschrift
legitimate 140 legitim
to lend 124 ausleihen
letter 68 hier: Buchstaben
level 10 Stufe, Niveau
liability 103 Haftung
liberalised 134 liberalisiert
licence 138 Lizenz
lifestyle 40 Lebensstil
to lift 134 heben, hieven
lighting 50 Beleuchtung
likely 118 wahrscheinlich
limited 57 begrenzt
to link; link 19 verbinden; Verbindung
listener 19 Zuhörer(in)
to live on welfare 114 von Sozialhilfe leben
loaf of bread 114 Laib Brot
local store 40,T Geschäft vor Ort
locally 135 vor Ort
locals 16 Einheimische
location 42 Lage, Standort
logical 17 logisch
logo 39 Logo, (Firmen)Emblem
loneliness 80 Einsamkeit
lonely 80,T einsam
long time no see 11 lange nicht gesehen
long-range 132,T Langstrecken-
long-term 80,T langfristig
to look forward to (-ing) 24 sich freuen auf (Schlussformel im Geschäftsbrief)
to look into 42 untersuchen, prüfen
to look like 15 aussehen wie
lounge area 113 hier: Halle
low-cost 30 preiswert
low-cost loans 124,T günstige Kredite
low-cost supplier 83 Billiglieferant
Lower House 72 Unterhaus
Luxembourg 27 Luxemburg
luxurious 130 luxuriös

M

machinery 64 Maschinen
machinist 136 Maschinenführer(in)
macroeconomic 82 gesamtwirtschaftlich, makroökonomisch
mail order 130 Versand(handel)
main 19 Haupt-
main course 114 Hauptgericht
mainstream 145 Hauptrichtung, Mainstream
to maintain 51 behaupten
maintenance 62 Wartung, Instandhaltung
major 42 bedeutend
major road 56 Hauptstraße
to make do 114 auskommen
to make one's living 126 seinen Lebensunterhalt verdienen
to make sense 46 einen Sinn ergeben
to make up one's mind 19 sich entschließen
maker 62 Hersteller(in)
male 11 männlich
Management Assistant in Event Organisation 99 Veranstaltungskaufmann, -frau
Management Assistant in Hotel and Hospitality 99 Hotelkaufmann, -frau
Management Assistant in Informatics 99 Informatikkaufmann, -frau
Management Assistant in Retail Business 99 Kaufmann, -frau im Einzelhandel
Management Assistant in Sports and Fitness 99 Sport- und Fitnesskaufmann, -frau
Management Assistant in Warehousing and Logistics 99 Fachkraft für Lagerwirtschaft
Managing Director 113 Geschäftsführer(in)
Manchester-based 98 in Manchester ansässig
mankind 58 Menschheit
to manufacture 51 herstellen
manufacturer 50 Hersteller(in)
margin 146 Rand, Seitenrand
marital status 88 Familienstand
mark 17 Zensur, Note
to mark down 34 (Preise, Noten) herabsetzen
market development 19 Marktentwicklung
Marketing and Export Department 113 Marketing und Exportabteilung
Marketing Manager 113 Abteilungsleiter(in) Marketing
marvellous 40,T fantastisch
Masala chicken 95 Hühnchen nach indischer Art
mashed potatoes 91 Kartoffelpüree
massive 122 massiv, wuchtig, enorm
mate 40,T Kollege, Kumpel

material 117 materiell
matter 32 Angelegenheit
maximum 126 maximal, Höchst-
mayor 30 Bürgermeister(in)
meat loaf 91 Hackbraten
mechanical engineering 10 Maschinenbau
medical 122 ärztlich, medizinisch
medical conditions 138 Krankheiten
Mediterranean (Sea) 122 Mittelmeer
medium-sized 46 mittelgroß
to meet the growing demand 50 der wachsenden Nachfrage gerecht werden
to melt 55 schmelzen
menu 114 Speisekarte
metal 14 Metall
middle-class 114 Mittelklasse
migration 122 (Ab-)Wanderung
millennium 30 Jahrtausend
Millennium Wheel 30 Millennium-Riesenrad
to mind one's own business 114 sich um seine eigenen Angelegenheiten kümmern
mining 56 Bergbau
misbehaviour 141 Fehlverhalten
misconduct 141 Fehlverhalten
to miss out on 116,T etwas verpassen
to misuse; misuse 114 missbrauchen; Missbrauch
mixed case 110 Groß- und Kleinschreibung
mobile phone 35 Handy
mobile phone operator 34 Mobilfunkbetreiber
mobilization 122 Mobilisierung
momentum 124,T Schwungkraft
to monitor 43 überwachen
moor 14 Moor
moral 118 Moral; moralisch
moreover 18 überdies, außerdem, zudem, ferner
mosque 144 Moschee
mother nature 124,T Mutter Natur
to motivate 130 motivieren
motorway 42 Autobahn
movie 121 Film
MP (Member of Parliament) 56 Mitglied des Parlaments
multiculturalism 144 Multikultur
multi-national corporation 134 multinationales Unternehmen
municipal 49 kommunal, städtisch
mutual 111 gegenseitig, gemeinsam
it's my turn 92 ich bin an der Reihe

N

nationality 24 Nationalität
native Americans 84 amerikanische Ureinwohner
native speaker competence 98 muttersprachleiche Kompetenz
native-born 86 gebürtig
nativity 116 Geburt, hier: Herkunft
natural 18 natürlich
nature 56 Natur
navigation system 42 Navigationssystem
neat 101 gepflegt
needle in a haystack 74 Nadel im Heuhaufen
neglect 30 Vernachlässigung
negotiator 96 Verhandlungspartner(in)
(the) Netherlands 27 Niederlande
netiquette 110 Umgangsformen im Internet
network 103 Netzwerk
nominal 137 nominal
non-polluting 57 umweltverträglich
non-returnable 50 Einweg-
non-smoking compartment 108 Nichtraucherabteil
non-stop 107 ohne Zwischenlandung
to note 135 bemerken, erwähnen, notieren
noun 18 Substantiv
novel 17 Roman
nucleus 138 Zellkern
nutrition 142 Ernährung

O

obesity 142 Fettleibigkeit
objective 145 Ziel
to observe 126 bemerken, feststellen
observer 141 Beobachter(in)
obsession 34 Besessenheit
to obtain 43 bekommen, erhalten
(to be) off sick 105 krankgeschrieben (sein)
offender 120 Straftäter(in)
Office Management Assistant 99 Bürokaufmann, -frau
office skills 10 Bürofertigkeiten
office worker 17 Büroangestellte(r)
officer 17 Beamte(r), Offizier(in)
officials 140 Vertreter
off-the-grid 62 nicht an das Versorgungsnetz angeschlossen
on the one hand 39 auf der einen Seite
on top of 48 zusätzlich zu
on your own 17 allein
once again 50 noch einmal
one way ticket 108 Fahrschein für einfache Fahrt
open air 114 Freiluft-, draußen
open plan office 113 Großraumbüro
to operate 48 verkehren, bedienen
operation 134 Fabrik, Betrieb
to opt out of 142 aus etwas aussteigen
ordinary 122 gewöhnlich
ore 56 Erz
organizer 14 Veranstalter
organs 78 Organe
ought to 51 sollte(n)
outback 69 Hinterland
outbound 31 ins Ausland
to outline 99 zusammenfassen
out-of-date 125 veraltet, überholt, altmodisch
outpouring 124,T Ausströmen
over and over again 54 immer wieder
overall grade 100 Gesamtnote
overcrowded 42 überfüllt
to overestimate 39 überschätzen
overhead projector 110 Tageslicht-Projektor
overseas 24 in Übersee
oversupply 135 Überangebot
overwhelming 142 überwältigend
ownership 103 Eigentum
ozone hole 50 Ozonloch
ozone layer 122 Ozonschicht

P

packaging 50 Verpackung
packing 11 Verpackung

painful 74 schmerzhaft
painter 88,T Maler(in)
pallet 128,T Palette
panel 62 Schalttafel
to panic; panic 118 in Panik geraten, in Panik versetzen; Panik
pants 91 (AE) lange Hose, (BE) Unterhose
paragraph 10 Absatz
paralysis 138 Lähmung
to paraphrase 101 umschreiben
parasailing 66 Fallschirmsegeln
parliament 27 Parlament
parliamentary leader 72 Parlamentsführer
to participate 32,T teilnehmen
participation 126 Beteiligung, Teilnahme
particulars 111 Einzelheiten
partnership 103 Partnerschaft , Personengesellschaft
part-time 10 Teilzeit
to pass away 122 entschlafen, hinscheiden; Umschreibung für: sterben
past tense 27 Vergangenheitsform
to pave 138 ebnen
to pay attention 122 aufmerksam sein
PDA (public demonstration of affection) 91 öffentliches Zuschaustellen von Zuneigung
peaceful 27 friedlich
peacekeeper 140 Blauhelmsoldat
peak time 46 Hauptverkehrszeit
peas 95 Erbsen
peat bog 64 Torfmoor
peer group 117 Gruppe von Gleichgesinnten
penal colony 69 Strafkolonie
pepper 91 Pfeffer
per 17 pro
per capita 137 pro Kopf
per-capita income 82 Pro-Kopf-Einkommen
to perceive 97 empfinden, wahrnehmen, erkennen
percentage 35 prozentualer Anteil
to perform 135 arbeiten, leisten, verrichten
performance 140 Leistung
permanent 30 ständig, permanent, dauerhaft
permanent position 98 Festanstellung
to permit 74 erlauben
pernicious 142 schädlich
personal property 66 persönliches Eigentum
personality 15 Persönlichkeit
personnel 140 Personal
pesticide 122 Pestizid
petrol station 43 Tankstelle
photographer 19 Fotograf(in)
photovoltaic (pv) 62 fotoelektrisch
phrase 16 Ausdruck
physical 125 physikalisch, physisch
to pick up 91 jmdn. oder etwas abholen
picker 128,T hier: Lagerarbeiter, der die bestellte Ware heraussucht
pie chart 71 Tortendiagramm
pin 138 Nadel
to pinpoint 42 genau anzeigen, genau bestimmen
planner 46 Planer(in)
plastic 49 Kunststoff, Plastik
plastic laminated material 124,T kunststoffbeschichtetes Material
platform 48,T Gleis
to play hard 79 sich kräftig amüsieren
to play truant 118 (die Schule) schwänzen
pleased 39 zufrieden, erfreut
pleasurable 78 angenehm
pleasure 71 Vergnügen
Pledge of Allegiance 92 Treueschwur (auf die Fahne)
plentiful 58 reichlich, häufig
plenty 58 viel, eine Menge
to plummet 135 einbrechen, stürzen
pocket money 11 Taschengeld
to poison; poison 122 vergiften; Gift
Poland 26 Polen
Polish 26 polnisch
polished 101 poliert
polite 40 höflich
political 27 politisch
poll 86 Umfrage, Wahl, Abstimmung,
polytechnic university (BE) 100 Fachhochschule
poorly paid 117 schlecht bezahlt
popularly elected 140 allgemein gewählt
to postpone 106 verschieben
potential 101 möglich
potential client 96 möglicher Kunde, mögliche Kundin
power line 62 (Stark-)Stromleitung
power plant 62 Kraftwerk
power station 16,T Elektrizitätswerk, Kraftwerk
powered 62 angetrieben, betrieben
practical 17 praktisch
Prague 30 Prag
to praise 138 loben
precious 126 wertvoll, kostbar
precise 54 genau
preconditions 137 Voraussetzungen
to predict 58 vorhersagen
preferences 72,T Vorlieben, Präferenzen
pregnant 116 schwanger
presentation 128,T Präsentation
pressed 101 gebügelt
previous 64,T vorhergehend
price swings 135 Preisschwankungen
primary energy consumption 64,T Primärenergieverbrauch
primary school 56 Grundschule
primary vote 72,T Erstabstimmung
principle 10 Grundsatz, Prinzip
privacy 70 Privatsphäre
pro-actively 98 mit Eigeninitiative
to proceed 48,T weitergehen, vorgehen
proceedings 96 Abläufe
process 27 Vorgang, Prozess
profession 100 Beruf
professional 101 professionell
to programme 43 programmieren
progress 46 Fortschritt
project 14 Projekt
projection 59 Prognose, Projektion
prolonged 97 anhaltend
to promote 34 fördern, werben für
promotion 34 (Verkaufs-)Förderung, Promotion, Beförderung
proof 126 Beweis
proper 34 korrekt, angemessen
properly 101 richtig
proposal 96 Angebot
proposed 72 vorgeschlagen
to prosecute 141 strafrechtlich verfolgen

prosecution 140 Strafverfolgung
prospective 99 zukünftig
prosperity 32 Wohlstand
to provide 46 sorgen für, liefern, bereitstellen
psychology 10 Psychologie
public library 56 Stadtbücherei
public service 9 öffentlicher Dienst
to publicize sth. 131 für etwas Werbung machen
to publish 88 veröffentlichen, publizieren
purchasing 111 Einkaufs-
purpose 71 Zweck
to push for 138 auf etw. drängen
to put (s.o./sth.) to the test 14 jmdn./etwas auf die Probe stellen
to put into practice 46 in die Praxis umsetzen, realisieren
to put the seal on 132,T besiegeln
puzzle 17 Rätsel

Q

qualification 10 Qualifikation, Befähigung
queue 128,T Schlange

R

race 101 Rasse
radioactive 57 radioaktiv
rail 46 Schiene
rampant 142 ungezügelt
range 50 Sortiment, Umfang, Reichweite
rare 126 selten
rate 31 Quote, Prozentsatz
rates of pay 98 Bezahlung
rave 78 Rave (größere Tanzveranstaltung zu Technomusik)
raw 54 Roh-, roh
raw materials 135 Rohstoffe
reaction 50 Reaktion
to realise/realize 30 feststellen, erkennen, bemerken
reality show 131 Fernsehsendung, die das tägliche Leben normaler Menschen zeigt
to rebel; rebel 118 rebellieren; Rebell
to recalculate 42 neu berechnen
recent 101 neueste
recently 103 kürzlich
reception desk 113 Rezeption, Empfangsschalter
receptionist 109 Rezeptionist(in)
to recharge 125 wieder aufladen
recipient 114 Empfänger
to reciprocate (gifts) 97 abwechselnd wirken, austauschen
to recognize 39 (wieder)erkennen, anerkennen
to recommend 68 empfehlen
recommendation 136 Empfehlung
to record 19 aufnehmen (Ton, Video etc.)
record-breaking 131 Rekord-
recording 56 Aufnahme
recruitment officer 98 Verantwortlicher für die Personalbeschaffung
recyclable 50 wiederverwertbar
to recycle 54 wiederverwerten
to rediscover 144 wieder entdecken
to reduce 41 verringern, senken
reduction 51 Verminderung
to refer to 40 sich beziehen auf
to refer to as 69 bezeichnen als
reference 24 Referenz, Hinweis
refreshments 112 Erfrischungen
refugee 88,T Flüchtling
to refuse 50 ablehnen, sich weigern
to register 68 registrieren
to regret 110 bedauern
to regulate 43 regulieren
to reignite 138 wieder entfachen,
to reject 106 zurückweisen
related 141 bezogen, verwandt
to release 64,T veröffentlichen, freilassen
religious 91 religiös
reluctantly 70 widerwillig
to rely on 43 sich verlassen auf
to remain 64,T bleiben
to remark 14 bemerken
remedy 82 Hilfsmittel
remote control 39 Fernbedienung
remote-controlled 128,T ferngesteuert
to render 134 hier: machen
renewable energy 62 erneuerbare Energie
repatriation 140 Rückführung
to replace 125 ersetzen
reported speech 56 indirekte Rede
representative 111 Vertreter(in)
reputation 132,T Ruf
request 111 Bitte, Anliegen, Anfrage
to require 24 benötigen, erfordern
research 62 Forschung
Research Manager 113 Abteilungsleiter(in) Forschung
reservations agent 98 Mitarbeiter(in) in der Reisereservierung
to reserve 26 reservieren
resources 64 Boden-, Naturschätze, Mittel
respectable 118 angesehen, anständig
responsibility 50 Verantwortung
responsible 24 verantwortlich
rest room (auch restroom) 92 Toilette (AE)
resurgence 64,T Wiederaufleben
retail sales 129 Einzelhandelsumsätze
retailer 50 Einzelhändler(in)
retailing 136 Einzelhandel
return ticket 108 Fahrschein für Hin- und Rückfahrt
returnable 50 Mehrweg-
to re-use 54 wiederverwenden
to reverse 142 Hier: ändern (Gewohnheit)
reward 142 Belohnung
to rewrite 43 neu schreiben, umschreiben
risk 18 Risiko
rival 132,T Konkurrent(in)
road works 43 Straßenbauarbeiten
role 56 Rolle
root 138 Wurzel
rose-tinted glasses 118 rosarote Brille
row 142 Streit, Krach
rubbish 50 Abfall, Müll
rude 97 unhöflich
to ruin 122 ruinieren
to run courses 10 Kurse anbieten
to run out of sth. 43 ausgehen, etwas nicht mehr haben
to run the business 130 das Geschäft führen
runaway 74 Ausreißer(in)
run-down 30 heruntergekommen

to rush 11 eilen
rush-hour 42 Hauptverkehrszeit
Russian 88,T russisch

S

to sacrifice; sacrifice 122 opfern; Opfer
safeguard 126 Schutzvorrichtung
salary 24 Gehalt
sales 18 Verkaufszahlen
sales department 105 Verkaufs-abteilung
salesman/saleswoman 17 Verkäufer(in)
salmon 114 Lachs
salutation 99 Anrede
sarcasm 110 Sarkasmus
satellite 42 Satellit
satisfaction 22 Zufriedenheit
satisfied 22 zufrieden
savings 88,T Ersparnisse
to scale 73 erklimmen
to scan 146 (Text nach Detailinformation) durchsuchen
schedule 92 Liste, Fahrplan, Stundenplan
scheduled flight 98 Linienflug
scheme 15 Projekt, Plan
scientist 62 Wissenschaftler(in)
scrap 54 Schrott, Fetzen
scrap yard 121 Schrottplatz
sea-crossing 69 Meeresüberquerung
seal 122 Seehund
seaside resort 56 Seebad
second hand 114 aus zweiter Hand, gebraucht
secondary modern school 102 Hauptschule
secondary school 69 weiterführende Schule, Hauptschule
secretarial staff 104 Büroangestellte
to secure 132,T sichern
to seek 98 suchen
selection 130 Auswahl, Sortiment
self-confidence 101 Selbstvertrauen
self-employed 18 selbstständig
self-esteem 74 Selbstachtung
selling tool 99 Verkaufswerkzeug
seminar 26 Seminar
senior 92 Schüler(in) der Abschlussklasse
sensor 43 Sensor, Fühler
to separate 54 trennen
separate 114 getrennt
separateness 144 Distanz
to service 18 (Auto, Maschine etc.) warten
service 91 Dienst, Dienstleistung, Gottesdienst
severe 79 schlimm, ernst, streng
sewage 122 Abwasser
sex 114 Geschlechtsverkehr, Geschlecht
sexual abuse 140 sexueller Missbrauch
shadow minister 72,T Schatten-minister
shaky 128,T unsicher
shame 73 Schande
to shape 143 bilden, formen
share 38 Anteil
sharp 59 scharf, deutlich, steil
to shift 54 verlagern
to shock; shock 118 schockieren; Schock
to shoe 69 beschlagen
shoplifting 118 Ladendiebstahl
shortage 135 Mangel
shorthand 34 Kurzschrift
short-term 86 kurzfristig
sick leave 22 krankheitsbedingte Abwesenheit
side salad 95 Salat zum Hauptgericht
sidewalk 92 Bürgersteig (AE)
sighting 74 Sichten
sights 15 Sehenswürdigkeiten
to sign 101 unterschreiben
to sign up for 114 sich einschreiben, anmelden
significant 58 bedeutend, bedeutsam
similar 24 ähnlich
similarity 93 Ähnlichkeit
to simplify 132,T vereinfachen
simply tailored 101 einfach geschnitten
sincerity 97 Ehrlichkeit
singer 17 Sänger(in)
single 27 einzig, hier: einheitlich
single room 109 Einzelzimmer
skilled 146 geschult
skills 98 Fähigkeiten
to skim 146 (Text) überfliegen
slight 58 leicht, schwach, gering
slogan 40 Slogan, Werbespruch
to slow down 27 verlangsamen
slow moving traffic 43 zäh fließender Verkehr
slum 140 Slum
slump 64,T Einbruch
to smash 120,T zerschlagen
smoker 14 Raucher(in)
smokestack 122 Schornstein
so-called 50 so genannt
soccer 92 Fußball (AE)
social studies 9 Sozialwissen-schaften
society 86 Gesellschaft
sociology 10 Soziologie
soft skills 101 Soziale Kompetenz
soil 125 Boden, Erde
solar power 58 Sonnenenergie
solicitor 72,T Rechtsanwalt, Rechtsanwältin
solid 103 fest, solide
sort 46 Art, Sorte
soup of the day 114 Tagessuppe
source 58 Quelle
Spain 27 Spanien
specialist 17 Spezialist(in)
species 122 (Tier-, Pflanzen-) Art, Spezies
spelling 34 Schreibweise, Rechtschreibung
sportsman 17 Sportler
sportswoman 17 Sportlerin
to spur 145 anspornen
stable 59 stabil, dauerhaft
stadium 128,T Stadion
staff 112 Belegschaft, Mitarbeiter
staffroom 14 Lehrerzimmer
stairs 113 Treppen
standard of living 58 Lebensstandard
to start off 78 anfangen, hier: seinen/ihren Ausgangspunkt haben
to start up (business) 19 (Geschäft) anfangen
starter 114 Vorspeise
state 27 Staat
to state 98 nennen
stationary 113 Bürobedarf
statistics 10 Statistik
steady 59 stabil, gleichbleibend
steering wheel 43 Lenkrad

step 24 Schritt, Stufe
to step up 124,T beschleunigen
stereotype 114 Stereotyp, Klischee
to stock 40,T auf Lager haben
stock exchange 126 Börse
stomach 91 Bauch
storage 113 Lager
storey 128,T Stockwerk, Etage
straight people 120,T normale Leute
stranger 126 Fremde(r)
straps 91 Träger
stream 122 Bach, Wasserlauf
to streamline 132,T rationalisieren
strength 30 Stärke
striking 134 auffallend
to struggle 135 sich abmühen
studies 9 Kurs, Studium
stuffy 136,T stickig
stumbling block 124,T Stolperstein
subheading 146 Zwischenüberschrift
subject title 110 Betreff
subsidy 135 Subvention
substantial 59 beträchtlich, erheblich
to substitute; substitute 54 ersetzen; Ersatz
substitution 125 Ersetzen, Austausch
subtitle 146 Untertitel
suburban 42 Vorort-
to suffer 34 leiden
to suggest 43 vorschlagen
suicide 118 Selbstmord
suit 101 Anzug, Kostüm
to suit 106 passen
suitable 17 geeignet
to sum up 18 zusammenfassen
summary 55 Zusammenfassung
sunchair 30 Liegestuhl
superb 40 erstklassig, großartig
superfast 40 sehr schnell
superior room 110 Luxuszimmer
supplier 54 Lieferant(in)
to supply 62 versorgen
to support 141 unterstützen
surface 56 Oberfläche
surplus 124,T überzählig
survey 22 Untersuchung, Umfrage
survival 114 Überleben
survivor 126 Überlebende(r)
sweets 128,T Süßigkeiten
to switch 39 umschalten, schalten

T

tactile 97 Körperkontakt mögen
to take a (language) test 26 einen Test machen, an einer (Sprach-) Prüfung teilnehmen
to take action 58 etwas unternehmen
to take care 43 aufpassen
to take out 68 herausnehmen
to take the view 51 der Ansicht sein
to take to pieces 55 zerlegen
to target 142 auf jmdn. abzielen (Werbung)
target group 40 Zielgruppe
task 24 Aufgabe
tax 56 Steuer
tax incentives 124,T Steueranreiz
tax-file number 66 Steuernummer
tax-system 82 Steuersystem
technological 42 technologisch, technisch
technologist 56 Techniker(in)
teenage 118 im Teenageralter
temperature 59 Temperatur
temporary employment 100 Anstellung mit Zeitvertrag
to tend 104 tendieren
tennis serve 39 Aufschlag (Tennis)
tense 19 Zeitform
term 92 Semester, Schulhalbjahr
to test 138 hier: überprüfen
to text 34 simsen, eine SMS-Nachricht senden
text message 34 SMS-Nachricht
thankfully 34 glücklicherweise
thanks to 46 dank
Thanksgiving 93 amerikanisches Erntedankfest
to the point 111 treffend, zur Sache gehörig
theory 86 Theorie
therapeutic cloning 138 therapeutisches Klonen
thief 126 Dieb(in)
third 128,T Drittel
thought 19 Gedanke
threat 122 (Be-)Drohung
to threaten 122 (be-)drohen
to throw out 116,T entlassen, feuern
thumb 34 Daumen
tiny 91 winzig
tip 114 Trinkgeld
tomato picker 83 Tomatenpflücker
ton 50 Tonne
topic 91 Thema
total 58 gesamt
total deliveries 132 Gesamtauslieferungen
touch 42 Berührung
tough 126 hier: schwierig, hart
tourist commission 68 Ausschuss für Tourismus
tourist guide 17 Fremdenführer(in)
town hall 48 Rathaus
trade barriers 82 Handelsbeschränkungen
trade union movement 72,T Gewerkschaftsbewegung
trading centre 30 Handelszentrum
traditional 62 traditionell, herkömmlich
traffic flow 43 Verkehrsfluss
traffic jam 42 Verkehrsstau
traineeship (BE) 98 Praktikum
tram 42 Straßenbahn
tramway 46 Straßenbahngleis
tranquilliser 116,T Beruhigungsmittel
to transfer 58 übertragen
travel agency 32,T Reisebüro
Travel Management Assistant 99 Reiseverkehrskaufmann, -frau
travel reservations 98 Reisebuchung
to treat 125 behandeln
treatment 122 Behandlung
treaty 27 Vertrag
trillion 58 Billion
to trivialise 138 etw. trivialisieren
trolley 128,T Einkaufswagen
troops 140 Truppen
tropical 122 tropisch
trunk 92 Kofferraum (AE)
trustworthiness 97 Treue, Vertrauen
to turn into 30 verwandeln, umwandeln in
to turn to 50 sich zuwenden
turnover 18 Umsatz
twin beds 110 zwei Einzelbetten
twin-engine 132,T mit 2 Motoren

U

ultimately 131 letztlich, schließlich
unattended 66 unbeaufsichtigt
unauthorized 82 unerlaubt
to undergo 138 durchmachen, sich unterziehen
underpaid 22 unterbezahlt
uneducated 114 ungebildet
unemployed 56 arbeitslos
unemployment benefit 86 Arbeitslosenunterstützung
unethical 138 unethisch
unforgettable 69 unvergesslich
unfulfilled 118 unerfüllt
to unify 132,T vereinheitlichen
uninhabitable 122 unbewohnbar
union activity 136 Gewerkschaftstätigkeit
union leader 136 Gewerkschaftsführer(in)
unique 32 einzigartig
United Kingdom 31 Vereinigtes Königreich
unity 27 Einheit
unless 122 wenn ... nicht, es sei denn
unlikely 54 unwahrscheinlich
unlimited company 18 Gesellschaft mit unbeschränkter Haftung
unpopular 46 unpopulär
unpractical 91 unpraktisch
unrivalled 103 unübertroffen
unsolicited 110 unerbeten
unstable 117 unstabil
unsung hero 140 ein Held von dem niemand spricht
unveiling 132,T Enthüllung
Upper House 72 Oberhaus
urgency 97 Dringlichkeit
urgent 83 dringend

V

vacancy 24 freie Stelle
vacation 92 Ferien, Urlaub (AE)
valet parking 109 Auto des Gastes wird eingeparkt
valid 66 gültig
valuable 101 wertvoll
value 39 Wert
variety 118 Auswahl
to vary 22 sich ändern, unterschiedlich sein
vast 114 riesig, weit
ventilation 136 Belüftung
to venture; venture 130 riskieren, sich wagen; Unternehmen
verdict 140 Urteil
veteran car 126 Oldtimer
via 48 über
vicious 140 teuflisch
victim 118 Opfer
video projector 110 Beamer
view 31 Blick, Sicht, Ansicht, Aussicht
viewer 39 Zuschauer(in)
viewpoint 39 Standpunkt
to violate 135 verletzen, missachten
violent 118 gewalttätig
virgin materials 124,T ungebrauchte Materialien
virtually 72 nahezu, fast
visa 88,T Visum
visual elements 70 Bildelemente
vital 135 entscheidend, lebenswichtig
vocational school (full-time) 102 Berufsfachschule
vocational school (part-time) 102 Berufsschule
voice recognition 43 Spracherkennung
voicemail 104 Anrufbeantworter
voluntary 140 freiwillig
voluntary organization 117 freiwillige Organisation
voter 72 Wähler(in)
voting system 72 Wahlsystem
voucher 142 Gutschein
vulnerable 135 verwundbar

W

wage 18 Lohn
to wane 124,T schwinden, abnehmen
wardrobe 92 Kleiderschrank
Warsaw 26 Warschau
wartime 122 Kriegszeit
waste 49 Müll, Abfall
waste management 54 Abfallentsorgung
wasteful 50 verschwenderisch
waxed paper 124,T Wachspapier
way of life 88 Lebensweise
to weaken 122 schwächen
weakness 101 Schwäche
wealth 58 Reichtum
wealthy 27 wohlhabend
wedding 19 Hochzeit
welfare 86 Wohl, Wohlfahrt, Sozialhilfe
well-paid 17 gut bezahlt
well-placed 143 gut platziert
westbound 48,T in westlicher Richtung
what ... like? 58 wie (ist er/sie/es) ...?
whatever 114 was auch immer
whereas 39 während
widebody 132,T Großraum-
willing 24 bereit
willingness 126 Bereitschaft
to wind up 140 enden, beenden
with reference to 24 mit Bezug auf
word-by-word 146 Wort für Wort
to work out 17 herausfinden, ausarbeiten, lösen
working class 22 Arbeiterklasse
working hours 22 Arbeitszeit
workmate 22 Arbeitskollege, -kollegin
workshop 14 Werkstatt, Arbeitsgruppen
World Bank 134 Weltbank
World Trade Organisation 83 Welthandelsorganisation
worldwide 39 weltweit
worried 86 besorgt
to worry 42 sich sorgen, sich beunruhigen
worship 91 Anbetung, Verehrung, Gottesdienst
to wrap 97 einwickeln
wrapper 142 Verpackung
wrinkles 66 Falten

Y

yearly 62 jährlich
Yours 15 Dein(e), Ihr(e), Liebe Grüße
Yours faithfully 24 Hochachtungsvoll
Yours sincerely 24 Mit freundlichen Grüßen
youth hostel 26 Jugendherberge
youth training place 118 Ausbildungsplatz für Jugendliche

Z

zinc 56 Zink

Basic vocabulary

AE = American English

A	
a couple of	ein paar
a little	etwas, ein bisschen
a lot of	viel
a number of	einige
to be able	können
abroad	im/ins Ausland
to accept	annehmen, akzeptieren
accident	Unfall
accurate	genau
activity	Aktivität
to add	hinzufügen, addieren
address	Adresse
advantage	Vorteil
to advertise	werben, inserieren
advertisement	Werbeanzeige, Anzeige
advertising	Werbung
advice	Rat, Ratschlag
to advise	(be)raten
to be afraid (of)	Angst haben (vor)
afternoon	Nachmittag
afterwards	danach
again	wieder
against	gegen
age	Alter, Zeitalter
ago	vor
to agree	einverstanden sein, zustimmen
agreement	Übereinstimmung, Übereinkunft
air	Luft
airport	Flughafen
all	alle(s), ganz
to allow	erlauben
almost	fast, beinahe
alone	allein
along	entlang
already	schon
also	auch
alternative	Alternative, Wahl; alternativ
although	obwohl
always	immer
among	zwischen, unter
amount	Betrag, Summe
angry	wütend
animal	Tier
to announce	bekannt geben, ansagen
announcement	Bekanntgabe, Durchsage
another	noch eine(r,s), ein(e) andere(r,s)
to answer; answer	(be)antworten; Antwort
any	irgendeine(r,s)
any time	jederzeit
anybody	irgendjemand
anyone	irgendjemand
anything	irgendetwas
to appear	(er)scheinen
apple	Apfel
to apply (for)	sich bewerben (um)
area	Gebiet
around	herum, um, ungefähr
arrival	Ankunft
to arrive	ankommen
article	Artikel
as	wie, da, als
as well	auch, ebensogut
to ask (for)	fragen, bitten (um)
at first	anfangs
at home	zu Hause
at last	schließlich, endlich
at least	wenigstens
at once	sofort
at present	im Augenblick, jetzt
at the moment	im Moment, jetzt
at work	bei der Arbeit
to attend (school)	(Schule) besuchen
attention	Aufmerksamkeit
aunt	Tante
autumn	Herbst
away	weg
awful	furchtbar, schrecklich

B	
back	Rücken; zurück
bad	schlecht, schlimm, böse
bag	Tüte, Tasche
baggage	Gepäck
to bake	backen
baker	Bäcker(in)
balcony	Balkon
bank	Bank (Geldinstitut)
basket	Korb
bathroom	Badezimmer
beach	Strand
beard	Bart
beautiful	schön
because	weil
because of	wegen
to become, (became, become)	werden
bed	Bett
bedroom	Schlafzimmer
before	vorher, ehe
to begin, (began, begun)	beginnen
beginning	Anfang
behind	hinter
to believe	glauben
to belong to	gehören
best	am besten
best wishes	alles Gute
better	besser
between	zwischen
bicycle	Fahrrad
big	groß
bike	Fahrrad (Kurzform)
bill	Rechnung
bird	Vogel

birth	Geburt
birthday	Geburtstag
bit	Stückchen
to bite, (bit, bitten)	beißen
black	schwarz
blackboard	(Wand-)Tafel
to blame (for)	verantwortlich machen (für)
block of flats	Wohnblock
blood	Blut
blue	blau
board	Brett
boat	Boot
body	Körper
to boil	kochen
to book; book	buchen; Buch
boot	Stiefel
border	Grenze
bored	gelangweilt
boring	langweilig
to borrow	sich ausleihen
both	beide
bottle	Flasche
bottom	Boden
box	Schachtel, Karton, Kiste
boy	Junge
boyfriend	Freund
bread	Brot
to break, (broke, broken); break	(zer)brechen; Pause
to break down	zusammenbrechen
breakfast	Frühstück
bridge	Brücke
to bring, (brought, brought)	bringen
to broadcast, (broadcast, broadcast)	übertragen, senden
brochure	Broschüre, Prospekt
brother	Bruder
brown	braun
to brush; brush	bürsten; Bürste
to build, (built, built)	bauen
building	Gebäude
to burn, (burnt, burnt)	brennen
business	Geschäft
busy	beschäftigt
but	aber
butcher	Metzger(in)
to buy, (bought, bought)	kaufen
buyer	Käufer(in)
by	durch, bei, neben, bis
by the way	übrigens
C	
cake	Kuchen
to calculate	(be)rechnen
to call; call	(an)rufen, nennen; Anruf
camera	Kamera
to camp; camp	zelten, lagern; Lager
can	können; Büchse, Dose
capital	Großbuchstabe, Hauptstadt
car	Auto

card	Karte
to care; care	sorgen; Sorge
career	Beruf, Laufbahn, Karriere
careful	sorgfältig
careless	sorglos
cargo	Fracht, Ladung
to carry	tragen
case	Fall, Koffer, Kiste
cash	Bargeld
cat	Katze
to catch, (caught, caught)	fangen
to cause; cause	verursachen; Ursache
ceiling	Zimmerdecke
centre	Zentrum, Mittelpunkt
century	Jahrhundert
certain	sicher
chair	Stuhl
chalk	Kreide
to change; change	wechseln, (sich) (ver)ändern; Wechsel(geld), Änderung
channel	Kanal
chart	Schaubild, Tabelle, Karte
cheap	billig
to check; check	kontrollieren; Kontrolle
cheers	Tschüs, zum Wohl
cheese	Käse
child	Kind
chips	Pommes Frites, Kartoffelchips
chocolate	Schokolade, Praline
choice	Wahl
to choose, (chose, chosen)	(aus)wählen
Christmas	Weihnachten
church	Kirche
cigarette	Zigarette
cinema	Kino
city	Stadt
class	Klasse
classmate	Klassenkamerad(in)
classroom	Klassenzimmer
to clean; clean	reinigen; sauber
to clear; clear	säubern, räumen; klar
clever	klug
to climb	klettern
clock	Uhr
to close; close	schließen; dicht, nahe
clothes	Kleidung
cloud	Wolke
coal	Kohle
coast	Küste
coat	Mantel
coffee	Kaffee
coin	Münze
cold	kalt
to collect	sammeln, abholen
collection	Sammlung
college	College, Fach(hoch)schule
colour	Farbe
to comb; comb	kämmen; Kamm
to come, (came, come)	kommen
comfortable	bequem
to communicate	kommunizieren, übermitteln
communication	Kommunikation, Übermittlung

company	Gesellschaft, Firma
to compare	vergleichen
competition	Wettbewerb, Konkurrenz
competitor	Konkurrent(in), Wettbewerbs-teilnehmer(in)
to complete; complete	vervollständigen; vollständig
to connect	verbinden
connection	Verbindung, Zusammenhang
to contact; contact	sich in Verbindung setzen mit; Verbindung, Kontakt
to contain	enthalten
container	Behälter
to continue	fortsetzen, weitermachen, weitergehen
to cook; cook	kochen; Koch, Köchin
cooker	Herd
to cooperate	zusammenarbeiten
cooperation	Zusammenarbeit
to copy; copy	kopieren, abschreiben; Kopie, Exemplar
corner	Ecke
to correct; correct	korrigieren, verbessern; richtig
correction	Verbesserung, Korrektur
to cost, (cost, cost); cost	kosten; Kosten
cotton	Baumwolle
to count	zählen, ins Gewicht fallen
counter	Ladentisch, Schalter
country	Land
couple	Ehepaar, Paar
course	Kurs, Lauf, Gang, Strecke
to cover; cover	(be)decken; Abdeckung, Umschlag
cow	Kuh
to cross	überqueren
crossing	Kreuzung
crowd	Menschenmenge, Menge
crowded	dicht gedrängt, voll
to cry; cry	schreien, weinen; Schrei
cup	Tasse, Pokal
curious	neugierig, seltsam
curtain	Vorhang
customer	Kunde, Kundin
customs	Zoll, Zoll-
to cut, (cut, cut); cut	schneiden; Schnitt
to cycle	Fahrrad fahren
D	
daily	täglich
to damage; damage	beschädigen; Schaden
to dance; dance	tanzen; Tanz
danger	Gefahr
dangerous	gefährlich
dark	dunkel
darkness	Dunkelheit
date	Datum, Zeitpunkt, Verabredung
date of birth	Geburtsdatum
daughter	Tochter
day	Tag
dead	tot
to deal, (dealt, dealt) (in, with); deal	handeln (mit, von); Geschäft

dear	lieb
death	Tod
to decide	(sich) entscheiden
decision	Entscheidung
deep	tief
to deliver	liefern, (Rede) halten
delivery	Lieferung
department	Abteilung
department store	Kaufhaus
to depend (on)	abhängen (von)
to describe	beschreiben
description	Beschreibung
to design; design	entwerfen, konstruieren; Entwurf, Muster
desk	Schreibtisch
to destroy	zerstören
destruction	Zerstörung
detail	Einzelheit, Detail
to develop	entwickeln
development	Entwicklung
to dial	(Telefon) wählen
to dictate	diktieren
dictation	Diktat
dictionary	Wörterbuch
to die	sterben
to differ (from)	sich unterscheiden (von)
difference	Unterschied
different	unterschiedlich
difficult	schwierig
difficulty	Schwierigkeit
dining room	Esszimmer
dinner	(Mittag-, Abend-)Essen
to direct (to); direct	richten, lenken (auf); direkt
direction	Richtung
dirty	schmutzig
disadvantage	Nachteil
to disagree (with)	anderer Meinung sein, nicht übereinstimmen mit
to discuss	besprechen, diskutieren
discussion	Gespräch, Unterredung, Diskussion
to dislike	nicht mögen
distance	Entfernung, Strecke, Abstand
doctor	Arzt, Ärztin
dog	Hund
door	Tür
down	hinunter, unten
downstairs	die Treppe hinunter, unten, im Erdgeschoss
to draw	zeichnen, ziehen
drawing	Zeichnung
to dream; dream	träumen; Traum
to dress; dress	(sich) anziehen; Kleid, Kleidung
to drink, (drank, drunk); drink	trinken; Getränk
to drive, (drove, driven); drive	fahren, (an)treiben; Fahrt
driver	Fahrer(in)
to dry; dry	trocknen; trocken
during	während
dustbin	Mülltonne
duty	Pflicht, Zoll

E	
each	jede(r,s)
each other	einander, sich
ear	Ohr
early	früh
to earn	verdienen
earth	Erde
east	Osten, östlich, Ost-, ostwärts
easy	leicht
to eat, (ate, eaten)	essen
economy	Wirtschaft
edge	Kante, Rand, Schneide
to educate	erziehen
education	Erziehung
to effect; effect	durchführen, erzielen, leisten; Wirkung
egg	Ei
either ... or	entweder ... oder
to elect	wählen
election	Wahl
electric	elektrisch
electricity	Elektrizität
electronics	Elektronik
else	sonst
emergency	Notfall
to employ	beschäftigen, einstellen
employed	beschäftigt
employee	Angestellte(r), Arbeitnehmer(in)
employer	Arbeitgeber(in)
employment	Beschäftigung, Anstellung, Arbeit
empty	leer
to end; end	(be)enden; Ende
energy	Energie, Kraft
engine	Motor, Triebwerk, Lokomotive
engineer	Ingenieur(in), Techniker(in)
engineering	Technik, technisch
to enjoy	genießen, sich freuen an
enjoyable	schön, angenehm, unterhaltsam
enough	genug
to enquire (about)	sich erkundigen (nach), fragen (nach)
enquiry	Anfrage, Erkundigung, Untersuchung
to enter	eintreten
entrance	Eintritt, Eingang
envelope	Umschlag
environment	Umwelt
equal	gleich
equipment	Ausrüstung, Ausstattung
especial	besondere(r,s)
even	sogar, selbst; eben, gleich
evening	Abend
event	Ereignis
ever	je
every	jede(r,s)
everybody	jeder
everyone	jeder
everything	alles
everywhere	überall
exact	genau
exam(ination)	Prüfung
example	Beispiel
excellent	ausgezeichnet, hervorragend
exciting	aufregend
to excuse; excuse	(sich) entschuldigen; Entschuldigung
exercise	Übung
to exist	existieren, bestehen
exit	Ausgang, Ausfahrt
to expect	erwarten
expensive	teuer
experience	Erfahrung
to explain	erklären
explanation	Erklärung
to export; export	exportieren, ausführen; Ausfuhr, Export
to express; express	ausdrücken, äußern; Schnellzug, als Eilsache
expression	Ausdruck
extra	besonders
eye	Auge

F	
to face; face	(einer Sache) ins Auge sehen, gegenübertreten; Gesicht
fact	Tatsache
factory	Fabrik
fair	Messe, Markt; gerecht, fair
to fall, (fell, fallen); fall	fallen; Fallen, Sturz; (AE) Herbst
false	falsch
family	Familie
famous	berühmt
far	weit
fare	Fahrpreis, Fahrgeld
farm	Bauernhof, Farm
farmer	Landwirt
farming	Landwirtschaft
fast	schnell
fat	Fett; dick
father	Vater
favourite	Lieblings-; Liebling
to feed, (fed, fed)	füttern
to feel, (felt, felt)	fühlen
feeling	Gefühl
few	wenige
field	Feld
to fight, (fought, fought); fight	kämpfen; Kampf
to fill	füllen
final	Finale, Endrunde; letzte(r,s), endgültig
to find, (found, found)	finden
fine	Strafe; fein, gut, schön
to finish; finish	beenden; Ziel, Vollendung
to fire; fire	feuern; Feuer
firm	Firma; fest, verbindlich
first	erste(r,s), zuerst
first of all	zuerst, vor allem
to fish; fish	fischen; Fisch

to fit; fit	(zusammen) passen, einbauen; gesund, in Form
to fix	befestigen, reparieren, besorgen
flat	Wohnung; flach
flight	Flug
floor	Boden
flower	Blume
to fly, (flew, flown); fly	fliegen; Fliege
to follow	folgen
food	Nahrung, Essen
foot	Fuß
football	Fußball
for example	zum Beispiel
foreign	ausländisch, fremd
forever	für immer
to forget, (forgot, forgotten)	vergessen
fork	Gabel
to form; form	formen, gestalten; Form, Formular, Klasse
fortunate	glücklich
forward(s)	vorwärts
to free; free	befreien; frei
freeway	(AE) Autobahn
frequent	häufig
fresh	frisch
fridge	Kühlschrank
friend	Freund(in)
friendly	freundlich
front	Vorderseite
fruit	Frucht, Früchte, Obst
full	voll
fun	Spaß
furniture	Möbel
further	weiter
future	Zukunft; zukünftig
G	
game	Spiel
garage	Autowerkstatt, Garage
garden	Garten
gate	Tor
gentleman	Herr
to get, (got, got)	bekommen, werden
to get married	heiraten
to get up	aufstehen
girl	Mädchen
girlfriend	Freundin
to give, (gave, given)	geben
to give up	aufgeben
glad	froh
glass	Glas
glasses	Brille
glove	Handschuh
to go, (went, gone)	gehen
to go down	untergehen, sinken, fallen
to go on	weitergehen, weitermachen
to go shopping	einkaufen
to go up	hinaufgehen, wachsen, steigen
good	gut
goodbye	auf Wiedersehen

goods	Güter, Waren
to govern	regieren
government	Regierung
grandfather	Großvater
grandmother	Großmutter
grandparents	Großeltern
grateful	dankbar
great	großartig, groß
green	grün
to greet	grüßen
grey	grau
grocer	Lebensmittelhändler(in)
ground	Boden
group	Gruppe
to grow, (grew, grown)	wachsen
grown-up	Erwachsene(r); erwachsen
growth	Wachstum
to guess	erraten
guest	Gast
H	
hair	Haar
half	Hälfte; halb
hall	Flur, Saal
ham	Schinken
handbag	Handtasche
handkerchief	Taschentuch
to hang, (hung, hung)	(auf)hängen
to happen	geschehen
happy	glücklich, heiter
harbour	Hafen
hard	hart, schwierig, anstrengend
hardly	kaum
hat	Hut
to hate	hassen
to have, (had, had)	haben, lassen
to have to	müssen
to head; head	anführen, fahren nach; Kopf, Leiter
headache	Kopfschmerz
headline	Überschrift
health	Gesundheit
healthy	gesund
to hear, (heard, heard)	hören
heart	Herz
to heat; heat	heizen; Hitze
heating	Heizung
heavy	schwer
height	Höhe
hello	Hallo
to help; help	helfen; Hilfe
helpful	hilfreich
here	hier
here you are	bitte(schön)!
to hesitate	zögern
Hi Tech (High Technology)	Hochtechnologie
to hide, (hid, hidden)	(sich) verstecken
high	hoch
hill	Hügel
history	Geschichte
to hit, (hit, hit); hit	schlagen; Schlag, Treffer, Erfolg

to hold, (held, held)	halten
hole	Loch
holiday	freier Tag, Feiertag, Urlaub
home	Wohnung, Heimat; zu/nach Hause
homework	Hausaufgabe
honest	ehrlich
to hope; hope	hoffen; Hoffnung
horse	Pferd
hospital	Krankenhaus
hot	heiß
hour	Stunde
house	Haus
housewife	Hausfrau
housework	Hausarbeit
how	wie
how are you?	wie geht es Dir/Ihnen?
however	jedoch
hundred	Hundert
hungry	hungrig
to hurry; hurry	sich beeilen; Eile
to hurt	verletzen
husband	Ehemann
I	
ice cream	Eis
idea	Idee
if	wenn, falls
ill	krank
illness	Krankheit
to imagine	sich vorstellen
immediate	unmittelbar, unverzüglich
immigrant	Einwanderer, Einwanderin
to immigrate	einwandern
to import; import	importieren, einführen; Einfuhr, Import
importance	Wichtigkeit
important	wichtig
impossible	unmöglich
to improve	(sich) verbessern
in fact	tatsächlich
in front of	vor
in italics	kursiv
in my opinion	meiner Meinung nach
in time	rechtzeitig
inch	(Maßeinheit) Zoll
to include	einschließen
income	Einkommen
to increase	zunehmen, steigen, erhöhen
indeed	tatsächlich
industry	Industrie
to inform	informieren
inside	in, innerhalb
instead (of)	anstatt
to instruct	unterrichten, anweisen
instruction	Unterricht, Anweisung, (Gebrauchs-)Anleitung
insurance	Versicherung
interest	Interesse, (Plural) Zinsen
interesting	interessant
to introduce	einführen, vorstellen

introduction	Einführung, Vorstellung
to invent	erfinden
invention	Erfindung
to invest	investieren
invitation	Einladung
to invite	einladen
island	Insel
J	
jacket	Jacke
jam	Marmelade
job	Arbeit, Stelle, Aufgabe
to join	(sich) anschließen, beitreten
journey	Reise
juice	Saft
to jump; jump	springen; Sprung
just	genau, soeben, nur
K	
to keep, (kept, kept)	(be)halten
key	Schlüssel
to kill	töten
kilogram(me)	Kilogramm
kind	Art, Sorte; freundlich
king	König
to kiss; kiss	küssen; Kuss
kitchen	Küche
knee	Knie
knife	Messer
to knock; knock	klopfen; Klopfen
to know, (knew, known)	wissen
knowledge	Wissen
L	
ladder	Leiter
lady	Dame
lake	See
lamp	Lampe
land	Land
landlady	Vermieterin, Wirtin
landlord	Vermieter, Wirt
language	Sprache
large	groß
to last; last	dauern; letzte(r,s)
late	spät
later on	später
latest	neueste(r,s)
to laugh	lachen
laughter	Gelächter
law	Gesetz
to lay, (laid, laid)	legen
lazy	faul
to lead, (led, led)	führen, leiten
leaflet	(Hand-)Zettel, Flugblatt
to learn, (learnt, learnt)	lernen
learner	Lernende(r), Anfänger(in)
least	wenigste(r,s), geringste(r,s)

to leave, (left, left)	verlassen
left	links
leg	Bein
leisure	Freizeit
lemon	Zitrone
to lend, (lent, lent)	verleihen
length	Länge
less	weniger
lesson	Unterrichtsstunde
to let, (let, let)	lassen
letter	Brief
to lie, (lay, lain)	liegen
to lie, (lied, lied)	lügen
life	Leben
to lift; lift	anheben, aufheben; Aufzug, Mitfahrgelegenheit
light	Licht, Lampe; hell, leicht
to like; like	mögen; wie
limit	Grenze, Beschränkung
line	Linie, Zeile
lip	Lippe
to list; list	aufführen, auflisten; Liste
to listen	zuhören
little	klein, wenig
to live; live	leben, wohnen; lebend, direkt, live
living room	Wohnzimmer
to load; load	laden; Ladung
local	lokal, ortsansässig, örtlich
long	lang
to look	schauen, (aus-)sehen
to look after	sich kümmern um
to look for	suchen
to look forward to	sich freuen auf
to look out	aufpassen, hinaussehen
to look up	nachschlagen, heraussuchen, aufblicken
lorry	Lastwagen
to lose, (lost, lost)	verlieren
loss	Verlust
lot	Menge, Los
lots of	viel
loud	laut
to love; love	lieben; Liebe
lovely	hübsch
low	niedrig
luck	Glück, Schicksal
lucky	glücklich (im Sinne von: Glück haben)
luggage	Gepäck
lunch	Mittagessen
M	
machine	Maschine
mad	verrückt
madam	gnädige Frau
magazine	Zeitschrift
mail	Post
main	Haupt-
mainly	hauptsächlich
to make, (made, made)	machen, herstellen

man	Mann
to manage	zu Stande bringen, schaffen, leiten, verwalten
management	(Geschäfts-)Leitung, Verwaltung, Durchführung
many	viele
map	(Land-)Karte
market	Markt
marmalade	(Orangen-)Marmelade
married	verheiratet
to marry	heiraten
mass	Masse
to match; match	zusammenbringen, passen; Wettkampf, Streichholz
mathematics	Mathematik
to matter; matter	etwas ausmachen; Angelegenheit
may	dürfen
maybe	vielleicht
meal	Mahlzeit
to mean, (meant, meant)	bedeuten, meinen
meaning	Bedeutung
(in the) meantime	inzwischen
meanwhile	inzwischen
meat	Fleisch
mechanic	Mechaniker(in)
media	Medien
to meet, (met, met)	(sich) treffen, kennenlernen
meeting	Treffen
member	Mitglied
to mend	reparieren
to mention	erwähnen
message	Mitteilung, Nachricht
method	Methode
metre	Meter
microwave	Mikrowelle
middle	Mitte
midnight	Mitternacht
might	Gewalt, Macht; könnte(n)
mile	Meile
milk	Milch
to mind; mind	etwas ausmachen; Meinung, Gedanken, Verstand
mine	Bergwerk
minute	Minute
mirror	Spiegel
Miss	Fräulein
to miss	vermissen, verfehlen, verpassen
mistake	Fehler
to mix; mix	mischen; Mischung
moment	Augenblick
money	Geld
month	Monat
moon	Mond
more	mehr
morning	Morgen
most	die meisten
mother	Mutter
mountain	Berg
mouth	Mund
to move	bewegen, umziehen
movement	Bewegung

Mr	Herr
Mrs	Frau (verheiratet)
Ms	Frau (verheiratet oder unverheiratet)
much	viel
music	Musik
must	müssen
must not	nicht dürfen
N	
to name; name	benennen; Name
narrow	eng
near	nahe
nearby	nahe gelegen
nearly	fast, beinahe
necessary	notwendig
to need; need	benötigen; Notwendigkeit
neighbour	Nachbar(in)
neither … nor	weder … noch
never	nie
new	neu
news	Nachrichten, Neuigkeit
newspaper	Zeitung
next	nächste(r,s)
nice	nett
night	Nacht
no	kein; nein
no longer	nicht mehr
no one	niemand
nobody	niemand
noise	Lärm
noisy	geräuschvoll
noon	Mittag
north	Norden; nördlich, Nord-, nordwärts
nose	Nase
not	nicht
not either	auch nicht
not even	nicht einmal
not yet	noch nicht
to note; note	notieren; Notiz
nothing	nichts
to notice; notice	bemerken; Aushang, Kenntnis, Beachtung
now	jetzt
nowadays	heutzutage
nowhere	nirgendwo
number	Zahl, Nummer, Anzahl
nurse	Krankenschwester
O	
o'clock	Uhr (Zeitangabe)
ocean	Meer, Ozean
of course	natürlich
off	von … weg
to offer; offer	anbieten; Angebot
office	Büro, Amt
often	oft
oil	Öl
old	alt
on time	pünktlich

once	einmal
only	nur
to open; open	öffnen; offen
opinion	Meinung
opposite	Gegensatz; gegensätzlich
or	oder
to order; order	bestellen, befehlen; Bestellung, Befehl
organization	Organisation
to organize	organisieren
other	andere(r,s)
out of work	arbeitslos
outside	draußen, außerhalb; Außenseite
to own; own	besitzen; eigene(r,s)
owner	Eigentümer(in)
P	
to pack; pack	packen; Packung
packet	Paket
page	Seite
to paint; paint	malen; Farbe
pair	Paar
paper	Papier
parents	Eltern
to park; park	parken; Park
part	Teil
to pass	vorbeigehen, (Prüfung) bestehen
passenger	Passagier, Fahrgast
passport	Pass
past	Vergangenheit
to pay, (paid, paid); pay	(be-)zahlen; Bezahlung
peace	Friede
pen	Füllfederhalter
pence	Plural von Penny
pencil	Bleistift
penny	Penny (engl. Münze)
people	Leute, Volk
per cent	Prozent
perfect	vollkommen, perfekt
perhaps	vielleicht
period	Periode, Zeitraum, Schulstunde
personal	persönlich
petrol	Benzin
to phone; phone	telefonieren; Telefon
photograph	Foto
to pick	pflücken, wählen
to pick up	nehmen, aufheben
picture	Bild
piece	Stück
pipe	Pfeife, Röhre
to place; place	stellen, legen; Platz
to plan; plan	planen; Plan
plane	Flugzeug
to plant; plant	pflanzen; Pflanze
plate	Teller
platform	Bahnsteig
to play; play	spielen; (Schau-)Spiel
player	Spieler(in)
playground	Spielplatz
please	bitte

pleasure	Vergnügen
pocket	Tasche
to point; point	zeigen; Punkt
police	Polizei
policeman	Polizist
policewoman	Polizistin
political	politisch
politician	Politiker(in)
politics	Politik
to pollute	verschmutzen
polluted	verschmutzt
pollution	Umweltverschmutzung
poor	arm
popular	beliebt, populär
population	Bevölkerung
port	Hafen
possibility	Möglichkeit
possible	möglich
post office	Post
postcard	Postkarte
postman	Briefträger
pot	Topf
potato	Kartoffel
pound	Pfund
power	Kraft
power station	Kraftwerk
practice	Übung
to practise	üben
to prefer	vorziehen
to prepare	vorbereiten
to present; present	schenken, vorstellen; Geschenk, Gegenwart; anwesend, gegenwärtig
to press; press	drücken, pressen; Presse
pretty	hübsch
price	Preis
pride	Stolz
prize	Preis, Gewinn
probable	wahrscheinlich
to produce	produzieren, herstellen
producer	Hersteller(in), Produzent(in)
product	Produkt
production	Herstellung, Produktion
to profit; profit	profitieren; Gewinn
profitable	rentabel, einträglich, lohnend
program(me)	Programm
to promise; promise	versprechen; Versprechen
to pronounce	aussprechen
pronunciation	Aussprache
to protect	schützen
protection	Schutz
proud	stolz
pub	Kneipe
public	Öffentlichkeit; öffentlich
to pull	ziehen
pupil	Schüler(in)
purpose	Zweck, Absicht
to push; push	schieben, stoßen; Stoß
to put, (put, put)	setzen, stellen, legen
to put on	aufsetzen
to put up	(Hand) heben, bauen, aufstellen

Q	
quality	Qualität
quantity	Menge
quarter	Viertel
queen	Königin
question	Frage
quick	schnell
quiet	ruhig
quite	ganz, ziemlich

R	
railway	Eisenbahn
to rain; rain	regnen; Regen
to raise	(hoch-)heben, erheben
rather	ziemlich
to reach; reach	erreichen; Reichweite
to read, (read, read)	lesen
reader	Leser(in)
ready	fertig, bereit
real	wirklich
reality	Wirklichkeit
reason	Grund
reasonable	vernünftig, günstig
to receive	erhalten, bekommen
recent	jüngst, kürzlich
reception	Empfang
receptionist	im Empfang arbeitende Person
to record; record	aufnehmen; Rekord, Aufnahme
red	rot
regular	regelmäßig
relative	Verwandte(r); relativ
to relax	(sich) entspannen
to remain	bleiben
to remember	sich erinnern an
to remind (of)	erinnern an
to rent; rent	mieten, vermieten; Miete
to repair; repair	reparieren; Reparatur
to repeat; repeat	wiederholen; Wiederholung
to replace	ersetzen
to reply; reply	antworten; Antwort
to report; report	berichten; Bericht
to rescue; rescue	retten; Rettung
to rest; rest	ruhen; Rest, Ruhe
to result; result	resultieren; Ergebnis
to return; return	zurückkommen; Rückkehr, Rückgabe
rich	reich
to ride, (rode, ridden); ride	reiten, fahren; Ritt, Fahrt
right	rechts
to ring, (rang, rung); ring	klingeln, anrufen; Ring, Anruf
to rise, (rose, risen); rise	(auf)steigen, zunehmen; Aufstieg, Zunahme
river	Fluss
road	Straße
roof	Dach
room	Zimmer
round	Runde; rund, herum
rubber	(Radier-)Gummi

rubbish	Abfall, Müll
to rule; rule	herrschen; Regel, Herrschaft
to run, (ran, run)	laufen
S	
sad	traurig
safe	sicher
safety	Sicherheit
sale	(Schluss-)Verkauf
salesman/woman	Verkäufer(in)
salt	Salz
same	gleiche(r,s), der-, die-, dasselbe
to save	retten
to say, (said, said)	sagen
school	Schule
sea	Meer
seat	Sitz
second	Sekunde; zweite(r,s)
secret	Geheimnis; geheim
secretary	Sekretär(in)
to see, (saw, seen)	sehen
to seem	scheinen
to sell, (sold, sold)	verkaufen
seller	Verkäufer(in)
to send, (sent, sent)	schicken, senden
to sentence; sentence	verurteilen; Satz, Urteil
serious	ernst
to serve	(be)dienen, servieren
to service; service	(Auto) warten; Dienst, Betrieb
to set, (set, set)	setzen, stellen, legen
several	mehrere
shall	sollen
to share; share	teilen; Anteil
sheet	Blatt, Bogen
shelf	Regal
to shine, (shone, shone)	scheinen, leuchten
to ship; ship	verschicken; Schiff
shirt	Hemd
shoe	Schuh
shop	Geschäft, Laden
shop assistant	Verkäufer(in)
short	kurz
should	sollte(n)
shoulder	Schulter
to shout; shout	rufen; Ruf
to show, (showed, shown); show	zeigen; Ausstellung, Vorstellung
to shut, (shut, shut)	schließen
sick	krank
side	Seite
sight	Sicht, Anblick, Sehenswürdigkeit
to sign; sign	unterschreiben; Zeichen
simple	einfach
since	seit
since then	seitdem
to sing, (sang, sung)	singen
sir	Herr (Anrede)
sister	Schwester
to sit, (sat, sat)	sitzen
to sit down	sich hinsetzen
size	Größe

skill	Fertigkeit, Fähigkeit, Können
skirt	Rock
sky	Himmel
to sleep, (slept, slept); sleep	schlafen; Schlaf
slim	schlank
slow	langsam
to slow down	verlangsamen
small	klein
to smell; smell	riechen; Geruch
to smile; smile	lächeln; Lächeln
to smoke; smoke	rauchen; Rauch
to snow; snow	schneien; Schnee
so far	bisher, bis jetzt
soap	Seife
soccer	Fußball
sock	Socke
soft	weich
solution	Lösung
to solve	lösen
some	einige
somebody	jemand
someone	jemand
something	etwas
sometimes	manchmal
son	Sohn
song	Lied
soon	bald
sorry	betrübt, tut mir Leid!, Entschuldigung!
to sound; sound	klingen; Geräusch, Klang
south	Süden, südlich, Süd-, südwärts
space	Raum, Platz, Weltraum
spare time	Freizeit
to speak, (spoke, spoken)	sprechen
speaker	Sprecher(in)
special	besondere(r,s), spezielle(r,s)
speech	Rede
to speed; speed	schnell fahren; Geschwindigkeit
to speed up	beschleunigen
to spell; spell	buchstabieren; Zauber(spruch)
to spend, (spent, spent)	ausgeben, verbringen
spoon	Löffel
spring	Frühling
staff	Belegschaft, Kollegium
stamp	Briefmarke
to stand, (stood, stood); stand	stehen, aushalten; Stand
to start; start	anfangen; Anfang
to state; state	erklären, darlegen; Staat, Zustand
station	Bahnhof
to stay	bleiben
still	dennoch, noch
stone	Stein
to stop; stop	(an)halten; Halt
to store; store	lagern, speichern; (AE) Laden, Kaufhaus, Lager
storm	Sturm
story	Geschichte
strange	merkwürdig
street	Straße

to strike; strike	streiken, schlagen; Streik
strong	stark
to study; study	studieren; Studium
subject	Fach, Thema
suburb	Vorort
to succeed (in)	Erfolg haben (in, bei)
success	Erfolg
successful	erfolgreich
such	solche(r,s)
sudden	plötzliche(r,s)
sugar	Zucker
to suit; suit	passen; Anzug
suitcase	Koffer
sum	Summe
summer	Sommer
sun	Sonne
sunshine	Sonnenschein
supermarket	Supermarkt
supper	Abendessen
to suppose	annehmen, vermuten
sure	sicher
to surprise; surprise	überraschen; Überraschung
surprised	überrascht
survey	Überblick, Umfrage
sweet	süß
to swim, (swam, swum)	schwimmen
to switch; switch	schalten, wechseln; Schalter
T	
table	Tisch, Tabelle
to take, (took, taken)	nehmen
to take part in	teilnehmen an
to take place	stattfinden
to talk; talk	sprechen, reden; Gespräch
tall	hoch(gewachsen), groß
to taste; taste	schmecken, probieren; Geschmack
tasty	lecker
tax	Steuer
tea	Tee
to teach, (taught, taught)	lehren, unterrichten
teacher	Lehrer(in)
team	Mannschaft
technical	technisch
technology	Technik, Technologie
telephone	Telefon
television (TV)	Fernsehen
to tell, (told, told)	erzählen
terrible	schrecklich
to test; test	testen, überprüfen, untersuchen; Klassenarbeit, Versuch
than	als
to thank	danken
that	dass, jene(r,s), welche(r,s)
then	dann, damals
there	dort
there is/are	es gibt
therefore	deshalb
these	diese
these days	heutzutage
thick	dick

thin	dünn
thing	Ding
to think, (thought, thought)	denken
thirsty	durstig
this	diese(r,s)
those	jene
though	obwohl, trotzdem
thousand	Tausend
through	durch
to throw, (threw, thrown)	werfen
thus	so
ticket	Fahrkarte
tidy	ordentlich, aufgeräumt
till	bis
time	Zeit
times	Mal(e)
tired	müde
tobacco	Tabak
today	heute
toe	Zeh
together	zusammen
toilet	Toilette
tomato	Tomate
tomorrow	morgen
tongue	Zunge
tonight	heute Abend
too	auch
tool	Werkzeug
tooth, teeth	Zahn, Zähne
top	Spitze
tour	(Rund-)Reise, Tour
toward(s)	in Richtung
tower	Turm
town	Stadt
toy	Spielzeug
to trade (in); trade	handeln (mit); Gewerbe, Handwerk, Handel
traditional	traditionell, herkömmlich
traffic	Verkehr
traffic light(s)	Ampel
to train; train	ausbilden, trainieren, eine Ausbildung machen; Zug
training	Ausbildung, Training
to translate	übersetzen
translation	Übersetzung
to transport; transport	transportieren; Transport
to travel	reisen, fahren
tree	Baum
trip	(Kurz-)Reise
trouble	Schwierigkeit(en)
trousers	Hose
true	wahr, richtig
truth	Wahrheit
to try; try	versuchen; Versuch
to turn	(sich) drehen, wenden
to turn left/right	links/rechts abbiegen
to turn on/off	ein-/ausschalten
to turn over	umdrehen, überschlagen
twice	zweimal
to type; type	tippen, (Maschine) schreiben; Art, Typ

typewriter	Schreibmaschine
typical	typisch
typist	Schreibkraft
U	
umbrella	Schirm
uncle	Onkel
to understand, (understood, understood)	verstehen
unemployed	arbeitslos
unemployment	Arbeitslosigkeit
unfortunate	unglücklich
unfortunately	leider
unfriendly	unfreundlich
unhappy	unglücklich
university	Universität
until (till)	bis
upon	auf
upstairs	(nach) oben, im Obergeschoss
to use; use	gebrauchen; Gebrauch
used	gebraucht, gewohnt
useful	nützlich
useless	nutzlos
usual	gewöhnlich
V	
vacation	(AE) Ferien
van	Lieferwagen
vegetable(s)	Gemüse
vehicle	Fahrzeug
very	sehr
village	Dorf
to visit; visit	besuchen; Besuch
visitor	Besucher(in)
vocabulary	Wortschatz, Vokabelverzeichnis
voice	Stimme
to vote; vote	abstimmen, wählen; Abstimmung, Stimme
W	
to wait	warten
waiter	Kellner
waitress	Kellnerin
to wake, (woke, woken) (up)	wecken, aufwachen
to walk; walk	gehen; Spaziergang
wall	Wand, Mauer
to want	wollen
war	Krieg
warehouse	Lager
to warm; warm	wärmen; warm
to warn	warnen
to wash	waschen
washing machine	Waschmaschine
to waste; waste	verschwenden; Abfall, Verschwendung
to watch; watch	beobachten, sehen; Uhr
water	Wasser
way	Weg, Art und Weise

weak	schwach
to wear, (wore, worn)	(Kleidung) tragen
weather	Wetter
week	Woche
weekend	Wochenende
weight	Gewicht
to welcome; welcome	willkommen heißen; Willkommen, Empfang
well	gut, gesund; na ja …
well-known	bekannt
west	Westen, westlich, West-, westwärts
wet	nass, feucht
what	was
what about …	was ist mit …
what else	was noch
what for	wofür, wozu
wheel	Rad
when	wann
where	wo
whether	ob
which	welche(r,s)
while	während
white	weiß
who	wer
whole	ganz
whom	wem, wen
whose	wessen
why	warum
wide	weit
wife	Ehefrau
will	Wille; werden
to win, (won, won)	gewinnen
window	Fenster
wine	Wein
to wish; wish	wünschen; Wunsch
within	innerhalb
without	ohne
woman	Frau
wonderful	wunderbar
wood	Holz, Wald
wool	Wolle
word	Wort
to work; work	arbeiten; Arbeit
worker	Arbeiter(in)
world	Welt
worse	schlechter
worst	am schlechtesten
worth	wert
to write, (wrote, written)	schreiben
writer	Verfasser(in), Schriftsteller(in)
wrong	falsch
Y	
year	Jahr
yellow	gelb
yesterday	gestern
yet	jedoch, schon
you're welcome	gern geschehen!
young	jung
youth	Jugend

Irregular verbs

Infinitive	Past tense	Past participle	German
awake	awoke	awoken	aufwachen
be	was, were	been	sein
beat	beat	beaten	schlagen
become	became	become	werden
begin	began	begun	anfangen, beginnen
bend	bent	bent	biegen
bet	bet	bet	wetten
bid	bid	bid	bieten
bite	bit	bitten	beißen
blow	blew	blown	blasen, wehen (Wind)
break	broke	broken	brechen
bring	brought	brought	bringen
broadcast	broadcast	broadcast	senden (Rundfunk)
build	built	built	bauen
burn	burned/ burnt	burned/ burnt	(ver)brennen
burst	burst	burst	platzen
buy	bought	bought	kaufen
cast	cast	cast	werfen, gießen
catch	caught	caught	fangen
choose	chose	chosen	wählen, auswählen
come	came	come	kommen
cost	cost	cost	kosten
cut	cut	cut	schneiden
deal	dealt	dealt	handeln mit, umgehen
dig	dug	dug	graben
do	did	done	tun, machen
draw	drew	drawn	ziehen, zeichnen
dream	dreamed/ dreamt	dreamed/ dreamt	träumen
drink	drank	drunk	trinken
drive	drove	driven	fahren
eat	ate	eaten	essen
fall	fell	fallen	fallen
feed	fed	fed	füttern
feel	felt	felt	fühlen
fight	fought	fought	kämpfen
find	found	found	finden
fly	flew	flown	fliegen
forget	forgot	forgotten	vergessen
forgive	forgave	forgiven	vergeben
freeze	froze	frozen	gefrieren
get	got	got	bekommen
give	gave	given	geben
go	went	gone	gehen
grind	ground	ground	mahlen, schleifen
grow	grew	grown	wachsen
hang	hung	hung	hängen
have	had	had	haben
hear	heard	heard	hören
hide	hid	hidden	verstecken

Infinitive	Past tense	Past participle	German
hit	hit	hit	schlagen, treffen
hold	held	held	halten
hurt	hurt	hurt	verletzen
keep	kept	kept	behalten
know	knew	known	wissen, kennen
lay	laid	laid	legen
lead	led	led	führen
learn	learned/ learnt	learned/ learnt	lernen
leave	left	left	lassen, verlassen
lend	lent	lent	(ver)leihen
let	let	let	lassen
lie	lay	lain	liegen
lose	lost	lost	verlieren
make	made	made	machen
mean	meant	meant	bedeuten
meet	met	met	treffen
pay	paid	paid	bezahlen
put	put	put	setzen, stellen, legen
read	read	read	lesen
ride	rode	ridden	reiten, fahren
ring	rang	rung	klingen, anrufen
rise	rose	risen	aufgehen, steigen
run	ran	run	rennen, laufen
say	said	said	sagen
see	saw	seen	sehen
seek	sought	sought	suchen
sell	sold	sold	verkaufen
send	sent	sent	schicken, senden
shake	shook	shaken	schütteln
show	showed	showed/ shown	zeigen
shut	shut	shut	schließen
smell	smelt	smelt	riechen
sing	sang	sung	singen
sit	sat	sat	sitzen
sleep	slept	slept	schlafen
speak	spoke	spoken	sprechen
spend	spent	spent	ausgeben
stand	stood	stood	stehen
swim	swam	swum	schwimmen
take	took	taken	nehmen
teach	taught	taught	lehren, unterrichten
tear	tore	torn	zerreißen
tell	told	told	sagen, erzählen
think	thought	thought	denken
throw	threw	thrown	werfen
understand	understood	understood	verstehen
wake	woke	woken	wecken
wear	wore	worn	tragen (v. Kleidung)
win	won	won	gewinnen
write	wrote	written	schreiben